Polymer Processing

This book covers polymer 3D printing through basics of technique and its implementation. It begins with the discussion on fundamentals of new-age printing, know-how of technology, methodology of printing, and product design perspectives. It includes aspects of CAD along with uses of Slicer software, image analysis software and MATLAB® programming in 3D printing of polymers. It covers choice of polymers for printing subject to their structure–property relationship, troubleshooting during printing, and possible uses of waste plastics and other waste materials.

Key Features

- Explores polymeric material printing and design.
- Provides information on the potential for the transformation and manufacturing, reuse and recycling of polymeric material.
- Includes comparison of 3D printing and injection moulding.
- Discusses CAD design and pertinent scaling-up process related to polymers.
- Offers basic strategies for improvement and troubleshooting of 3D printing.

This book is aimed at professionals and graduate students in polymer and mechanical engineering and materials science and engineering.

Emerging Materials and Technologies
Series Editor: Boris I. Kharissov

The *Emerging Materials and Technologies* series is devoted to highlighting publications centered on emerging advanced materials and novel technologies. Attention is paid to those newly discovered or applied materials with potential to solve pressing societal problems and improve quality of life, corresponding to environmental protection, medicine, communications, energy, transportation, advanced manufacturing, and related areas.

The series takes into account that, under present strong demands for energy, material, and cost savings, as well as heavy contamination problems and worldwide pandemic conditions, the area of emerging materials and related scalable technologies is a highly interdisciplinary field, with the need for researchers, professionals, and academics across the spectrum of engineering and technological disciplines. The main objective of this book series is to attract more attention to these materials and technologies and invite conversation among the international R&D community.

Nanocosmetics
Drug Delivery Approaches, Applications and Regulatory Aspects
Edited by Prashant Kesharwani and Sunil Kumar Dubey

Sustainability of Green and Eco-friendly Composites
Edited by Sumit Gupta, Vijay Chaudhary, and Pallav Gupta

Assessment of Polymeric Materials for Biomedical Applications
Edited by Vijay Chaudhary, Sumit Gupta, Pallav Gupta, and Partha Pratim Das

Nanomaterials for Sustainable Energy Applications
Edited by Piyush Kumar Sonkar and Vellaichamy Ganesan

Materials Science to Combat COVID-19
Edited by Neeraj Dwivedi and Avanish Kumar Srivastava

Two-Dimensional Nanomaterials for Fire-Safe Polymers
Yuan Hu and Xin Wang

3D Printing and Bioprinting for Pharmaceutical and Medical Applications
Edited by Jose Luis Pedraz Muñoz, Laura Saenz del Burgo Martínez, Gustavo Puras Ochoa, Jon Zarate Sesma

Polymer Processing
Design, Printing and Applications of Multi-Dimensional Techniques
Abhijit Bandyopadhyay and Rahul Chatterjee

For more information about this series, please visit: www.routledge.com/Emerging-Materials-and-Technologies/book-series/CRCEMT

Polymer Processing

Design, Printing and Applications
of Multi-Dimensional Techniques

Abhijit Bandyopadhyay and Rahul Chatterjee

CRC Press
Taylor & Francis Group
Boca Raton London New York

CRC Press is an imprint of the
Taylor & Francis Group, an **informa** business

First edition published 2024
by CRC Press
6000 Broken Sound Parkway NW, Suite 300, Boca Raton, FL 33487–2742

and by CRC Press
4 Park Square, Milton Park, Abingdon, Oxon, OX14 4RN

CRC Press is an imprint of Taylor & Francis Group, LLC

ISBN: 9781032393483 (hbk)
ISBN: 9781032393506 (pbk)
ISBN: 9781003349341 (ebk)

DOI: 10.1201/9781003349341

Typeset in Times
by Apex CoVantage, LLC

Contents

About the Authors

Prof. Abhijit Bandyopadhyay is presently working as a full professor in the Department of Polymer Science and Technology, University of Calcutta. He is also acting as the Technical Director in the Board of Directors of South Asia Rubber and Polymers Park (SARPOL), West Bengal, India. Prof. Bandyopadhyay has published 110 papers in high-impact international journals, authored 5 books and has filed 2 Indian patents so far. He is the Fellow of International Congress for Environmental Research, Associate Member of Indian Institute of Chemical Engineers and Life Members of Society for Polymer Science, Kolkata Chapter and Indian Rubber Institute. His research areas include polymer blend and composites, polymer nanocomposites, reactive blending, hyperbranched polymers, polymer hydrogels, waste-polymer composites and polymer 3D printing using fused deposition modelling.

Mr. Rahul Chatterjee has a B.Tech in mechanical engineering and M.Tech in mechatronics engineering. He is currently pursuing doctoral research from the Department of Polymer Science & Technology at the University of Calcutta, India. His research area includes CAD design and fused deposition modelling with thermoplastic and thermoplastic elastomeric materials. He has published three papers in peer-reviewed international journals and is currently undergoing an internship program at Hari Shankar Singhania Elastomer and Tyre Research Institute (HASETRI), JK Tyre, Mysuru, India in the tyre design and simulation group.

1 Fundamentals of New-Age Printing

1.1. CONCEPT OF ADDITIVE MANUFACTURING AND ITS TYPES

Creative and cutting-edge research efforts linked with production processes, materials, and product design are critical to the evolution of industries. In addition to the traditional expectations for cheap prices and high quality, market rivalry in today's manufacturing industries is associated to demands for products that are complex, have shorter life cycles, have faster delivery times, require customization, and require less qualified personnel [1]. Due to its numerous advantages, additive manufacturing has become a current trend in manufacturing processes [2]. Because AM is most similar to "bottom up" production, where a structure can be constructed into its intended configuration using a "layer-by-layer" foundation as opposed to casting or forming utilising technologies like forging or machining, it differs fundamentally from conventional formative or subtractive manufacturing [3].

There are many advantages and disadvantages to choosing an additive or subtractive process for manufacturing a component. Before delving into them, let's take a look at the main contrasts between the two strategies [4]. Subtractive composition involves removing unwanted material from most of the material. The creation of additives is represented by the accumulation of parts using only the required materials. Subtractive technology is a technique that involves removing layers of materials to get a specific form. Subtractive technologies have evolved dramatically in the previous 20 years. Traditional code creation, such as G and M codes, has been supplanted by three-dimensional (3D) extremely complicated surface modelling software [1].

Both techniques are like the front and back of the same coin, both of which make it easy to create and use and but they don't have to be the same. Both are accurate

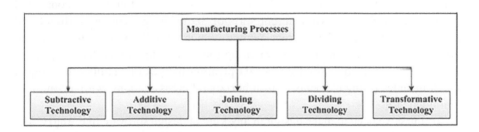

FIGURE 1.1 Classification of Manufacturing Process.

Source: Reprinted with permission from [1]

DOI: 10.1201/9781003349341-1

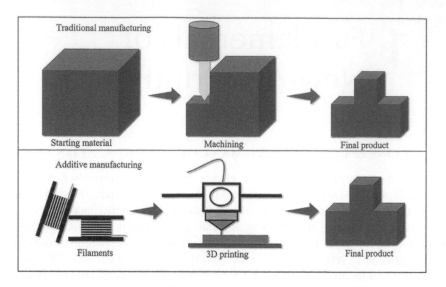

FIGURE1.2 Conceptual Representation of Subtractive Manufacturing vs Additive Manufacturing.

Source: Reprinted with permission from [5]

enough for prototyping, but simple enough for a learning environment. Both methods have their pros and cons. As a general recommendation, it is difficult to describe one type for the other. Recommendations really depend on the application and the desired outcome. If the desired result is a faster product and the ability to produce multiple parts at once, AM is the most beneficial path [6].

1.2. WHAT IS 3D PRINTING?

The most effective technology that is altering the current manufacturing environment is additive manufacturing [7]. In especially for designing complex parts, polymer 3D [8] printing is fast displacing traditional polymer processing methods worldwide. A 3D file source or a three dimensional design [9] is typically split into a series of layers, each of which generates a set of computer-controlled instructions. There are two different sorts of 3D printing techniques: direct and indirect. In contrast to the indirect method, which uses 3D printing to create the model, direct 3D printing involves creating the design directly on the 3D printer. In order to manufacture solid items, 3D printing merges the additive technique, which includes piling elements in thin horizontal cross-sections, with a computer command [10].

So at first we need to make design in any kind of design software like Auto CAD, SolidWorks, Maya then save the file in.stl extension then convert it into G-code after that load it into micro SD card and print from that (Figure 1.3). A level of standards that classify the multiple procedures were developed in order to remove any

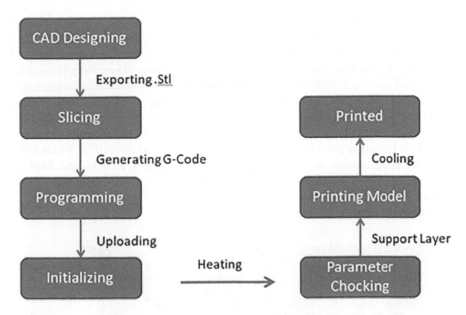

FIGURE 1.3 Block Diagram of 3D Printing From CAD Design to Printing.

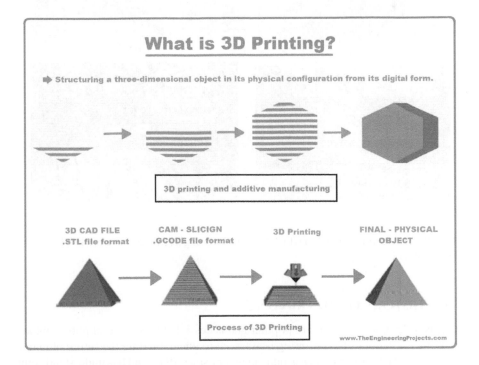

FIGURE 1.4 3D Printing Process.

Source: Reprinted with permission from [12]

ambiguity regarding the various forms of additive manufacturing. The International Organization for Standardization [11] developed these categories in 2010 and gave additive manufacturing techniques precise definitions, so there are seven types of techniques are available in present era. In many industries, 3D printing is used for full-fledged manufacturing. That is a significant leap. And, as a result of the increased accessibility, the trends have taken root. Although some metals can also be used for 3D printing, but the most common material utilized is plastic polymer. Plastics, on the opposite hand, have the benefit of being lighter than their metallic counterparts. This is mainly great in industries like automobile and aircraft, in which lightweighting is a situation and better gasoline performance can be achieved. Medical applications require a wide range of material qualities, which can be achieved with polymer 3D printing.

So here are some 3D printed objects.

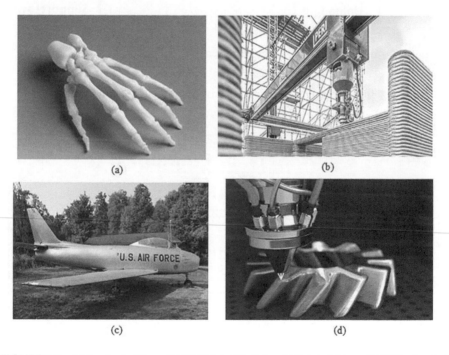

(a)   (b)

(c)   (d)

FIGURE 1.5   (a) Implanted Bone (b) 3D Printed Building (c) Printed Aircraft (d) Mechanical Gears From FDM.

1.3.   HISTORY OF 3D PRINTING

As this technology grows more popular and accessible to the general public, understanding the history of 3D printing is essential to understanding the future of manufacturing. 3D printing has been around in theory since 1945, and has understand come into practice (although primitively) since 1971, promising a faster and more efficient technique of producing goods. This technology has been studied for about 80 years, but has come into practice in the recent 40 years of development and appears to be spanking new [13]. Hideo Kodama, a Japanese innovator was the pioneer to introduce 3D printing using the additive technique in 1981. He invented a product that hardens polymers and turns them into solid objects using ultraviolet light. This is a prerequisite for stereolithography (SLA). Charles Hull pioneered stereolithography, a procedure equivalent to 3D printing that employs technology to generate miniature versions of products for testing before investing time and money in creating the final product. Layer by layer, the object is printed, washed with a solvent, then hardened with UV radiation. The 3D models are created using computer-aided design (CAD). Over the last 30 years, 3D printing has evolved and developed. SLA, SLS, and FDM depict the history of 3D printing and how it evolved into a critical manufacturing tool [13]. Hideo Kodama (Nagoya Municipal Industrial Research Institute) devised and documented the first two additive manufacturing processes based on photo-hardening of polymeric profiles between 1980 and 1981 [14]. Both photopolymerization and stereolithography can be traced back to this fundamental study. The inventor filed a patent application, but did not follow up within the statutory one-year period after filing. The SLA-1, 3D Systems' first commercialized rapid prototyping system, was unveiled in 1987, and the first of these systems was commercialized in 1988 [15]. S. Scott Crump developed and patented fused deposition modelling in 1989–1990, making it the most popular 3D printing technology to date, particularly for amateurs and moderate labs (FDM). It entails layer-by-layer deposition of fused material, most frequently plastic, using an STL file [14]. In the year 1993, MIT invented what is now known as same as 3D printing. The method involved layer-by-layer binding of a powder particles using an inkjet printer. The selective laser sintering technology was invented in 1995 by the Fraunhofer Institute ILT in Aachen. Because metal alloys are used, the method produces precise and structurally strong outputs and can manage multilayered and complicated designs. In 1990 Wake Forest Institute for Regenerative Medicine successfully demonstrated bioprinting techniques. In an effort to disseminate and democratise AM technology, Adrian Bowyer created the RepRap open-source project in 2004, which aims to create self-replicable 3D printers. Shapeways makes its debut in the Netherlands in 2008. It is a web-based service that allows customers to send 3D files to have products printed and delivered to the specified destination. The service use a variety of procedures and materials, including several precious metals today. In 2009 Makerbot released a 3D printer DIY kit, which will greatly aid in the widespread adoption of the technology in many homes. The Southampton University Laser Sintered Aircraft (SULSA), an unmanned aerial vehicle whose framework was printed with a pixel density of 100 micrometres per layer by a laser sintering device from the wings to the essential control surfaces, demonstrated the possibilities offered by 3D printing techniques as production instead of pure prototyping tools in 2011. Using "snap fit" procedures, the unmanned aerial vehicle (UAV) may be constructed without the use of any tools.

Airbus Operation GmbH filed a patent in 2014 to print a whole aeroplane construction in 3D mode. The technology is especially intriguing because it has 4D printing-like characteristics [14, 15].

Nowadays, direct 3D printing for biomedical parts is now the most used method, and the medical industry has adopted it quickly. In today's era lots of work has been also done to flourish this medical field like Ferretti et al. [16], which was chosen TPU because of its unusual qualities, which provide high strength and flexibility while also allowing for speedier printing. The study's goal was to see how effective TPU in FDM was at manufacturing biological goods. Tzounis et al. [17] used FDM technique to print and sonochemistry thin-film deposition technique to create antimicrobial surgery equipment. A surgical retractor was constructed using a typical polylactic acid (PLA) thermoplastic filament, and a simple and scalable sonochemical deposition of a thin coating of silver (Ag) nanoparticles (NPs) was achieved. According to Ballard et al. [18], additive manufacturing (AM) technology allows antibiotics to be incorporated into 3D printed structures. Antibiotics were present in medical implants, prostheses, as well as procedural and surgical tools. They were antibiotic-filled PLA pellets that were extruded as filaments of the required dimensions to increase surface area for drug distribution. Pfahn et al. [19] created 3D printed thermoplastic sponge sticks, which might be used to make towel clamps, scalpel handles, toothed forceps, and other items. Two types of printing materials were used to create the material: ABSplus-P430 plastic and ABSplus-P430 paper.

As a result of this technology, we can build nearly anything just by creating a computer file. Rankin et al. [20] used FDM to make surgical implants, tissue scaffolds, and organs. They'd also made a PLA-based surgical retractor prototype for the Army and Navy. The retractor was sterilised using FDA-approved glutaraldehyde procedures before being tested for bacteria using a polymerase chain reaction. Fresh PLA from the 3D printer was sterile, and no polymerase chain reaction result was produced. In 2019, Singh et al. [21] developed biomedical equipment for spine treatment and orthopaedic instruments and devices using a novel polymer called poly ether ether ketone, or PEEK. Shetty et al. [22] 3D printed two hard-tissue surgical models for orthopaedic, pelvic, and craniofacial procedures with great success. PLA and ABS were used for the sculptures due to vary their mechanical characteristics and biocompatibility. Not only the medical field, but other sectors like aerospace defense also adopted this technology quickly; some works already done by different researchers are as follows: C. Joshi et al. [23] was investigating the stock of popular 3D printing techniques in aircraft and the advantages of these technologies over standard production procedures are discussed. Depending on the AM technology employed and the physical state of the material, polymers, ceramic composites, aluminium, steel, and titanium objects can be printed with a minimum layer thickness of 20 to 100 m. However, in the aerospace industry, titanium- and nickel-based alloys are more important. In 2020, Kalender et al. [24] was researched about the materials which used in aerospace industry. Most of the time they used ABS and PLA filaments for making the model, PET, PVA, HIPS, Nylon, Carbon, and glass-reinforced filaments are among the other materials that can be employed. Durable filaments like wood, bark, bronze, copper, and carbon fibre blended with PLA plastic are another form of filament. These materials are commonly mixed at a 30–40% proportion with PLA. In present time aerospace and defense sectors using the AM technology for creating models, test unit production and prototyping, fitment parts, visualization parts, tooling [25], replacement parts and for the building components for aerospace and defense system. But in future they will also try to print Embedding additively manufactured electronics parts directly; they also want to print the repair parts on the battlefield, regenerative medicine to treat severe battlefield injuries, want to print structures using light-weight and high-strength materials with minimal waste. Parts made by additive manufacturing are used by BAE Systems, this is the largest defence manufacturer in the UK, to maintain and service Royal Air Force aircraft. BAE Systems could lower Royal Air Force servicing and maintenance expenses by implementing AM [25].

Ultimately, while the primary additive manufacturing techniques' founding ideas date back to the 1980s, additional development and interaction among techniques has steadily reflected a shift in their possible usage.

1.4.   PRESENT SCENARIO 3D PRINTING

The market for 3D printing is dominated by the printers' category. It mostly consists of desktop and industrial printers. Due to its many benefits over conventional production techniques, 3D printing has been extremely popular during the past ten years. A number of benefits are provided by the production process, including design flexibility, quick prototyping, print-on-demand, minimum waste, quick designing and production, accessibility, and time and cost efficiency. During the projected period, the market for commercial 3D printers is anticipated to expand more rapidly. Such printers can be employed for domestic, commercial, and industrial printing needs. These printers can be configured to print any component or module continuously, lowering material costs and assuring minimal or no waste [26]. When materials need to be lighter while maintaining their strength, 3D printing is a godsend, as a result [27]. In terms of adopting 3D manufacturing techniques, the medical and dentistry industries are currently leading the way, along with the aerospace industry. The printing of tissues, organs, implants, prostheses, and other medical items is currently done in the medical sector using 3D manufacturing technology. Using desktop printers to quickly generate complicated automotive parts, 3D printing technology is mostly employed in the automotive industry for prototype development. 3D printing is great for meshing designs from the cloud of 60,000 users to generate a joint byproduct for a company like Local Motors [28], which is deeply ingrained in its modular production process. Consider how the frame is conceived by an Italian, the engine is designed by a Michigan native, the wheels are made by a third party in China, and the final product is printed in a number of micro-factories before being assembled [29]. Pîrjan and Petroşanu done extensive research on additive manufacturing in 2013 [30] and discovered that the use of 3D printing technology in the automotive sector has been hampered by a number of issues, including the high cost of procuring 3D technology materials and resource constraints in industrial equipment. Despite these obstacles, Dwivedi, Srivastava, and Srivastava [30] reported in 2017 that 3D technology has been widely used by the automobile sector and has greatly transformed the way of making vehicles in today's society. Nonetheless, there is a scarcity of public data on the influence of 3D technology in the automotive industry. After overcoming all of the obstacles, Sreehita et al. [31] confirmed the existence of an electric car manufactured utilising 3D printing technology in 2017. The ABS carbon-filter substance, also known as strati, was employed by the 3D printing machinery to construct the components of the battery powered automobile in 44 hours by the company named as local motors, this study demonstrates how 3D printing technology may drastically cut the time it takes to construct a vehicle while also lowering the manpower necessary. As a result, automotive manufacturers may save a lot of money. Buildings will also be added in this new era of printing, however, due to the lack of formwork that temporarily supports the element while it sets, 3D-printable concrete requires specialised features, especially in the fresh condition, such as reduced slump and rapid hardening. Because a printed combination with reduced slump must also have some self-compaction, these two seemingly contradictory criteria create the problem of creating a printable mixture that achieves both [32]. Firstly, the materials must be extrudable and retain its geometry once deposited on the print bed; next, the printed layers should not crumble under the load of successive layers; and after that, superior qualities are attained in the hardened state by assuring the bond strength between layers [33]. A good mix should be formed of binder material and fine sand particles, as well as mineral or chemical modifiers, to ensure extrudability and geometry retention of a fresh cementitious material. Ordinary Portland cement (C), fly ash (FA), granulated blast furnace slag, silica fume (SF), and nano-silica are the most frequent components in the binder portion of concrete, which are mixed in various proportions. FA is a by-product that is obtained as trash from a coal-fired power plant. Class F FA has a silica percentage of more than 70%, while Class C FA has a CaO content of 15–30%. Class C FA has distinct self-hardening characteristics due to its increased CaO concentration [34].

1.4.1. Estimated Growth of 3D Printing in Automotives from 2012 to 2025

The marketplace for automobile 3D printing technology is anticipated to reach $2391 million by 2022, up from $621 million in 2015, according to Automotive 3D printing Segment by New Tech and Material-Global Opportunity Analysis and Industry Forecast, 2015–2022. From 2016 to 2022, the market is anticipated to grow at a rate of 21.8%. The following figure summarizes how 3D printing methods have significantly increased in the automobile industry when compared to other fields, per Lux Research, Inc. [28].

As organizations hunt for methods to acquire a competitive advantage, the pace of innovation is quickening across industries. Meanwhile, availability and labour shortages are driving the adoption of digital technologies in the design and development phase to accelerate the time to introduce for new products. So we are seeing that people are becoming more aware of additive manufacturing's ability to help the environment by lowering energy use and trash. When compared to CNC machining, where components are chopped away and discarded, additive manufacturing can reduce material prices and waste by over 90% while also cutting energy consumption by 25% to 50%. Furthermore, because parts can be manufactured closer to where they are needed, transportation costs are decreased, saving fuel and lowering carbon footprints. Furthermore, we have seen a growing interest in lightweight, additively built polymer parts to enhance fuel economy without sacrificing durability and dependability in mobility industries such as aerospace and electric cars. To give you an example, every pound a jet loses saves 14,000 gallons of gasoline over the course of a year [35].

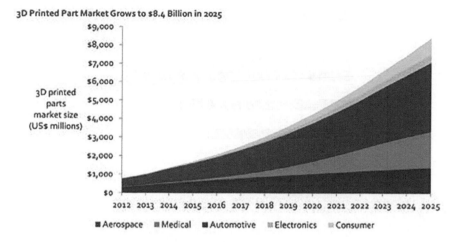

3D Printed Part Market Grows to $8.4 Billion in 2025

■ Aerospace ■ Medical ■ Automotive ■ Electronics ■ Consumer

FIGURE 1.6   3D Printed Part Market.

Source: Reprinted from Lux Research. Inc [28]

1.4.2. ACCESSIBLE 3D PRINTING MATERIAL OPTIONS ARE SURGING

Since 2019, there has been a sharp rise in the variety of 3D printing materials that industries are using. Whereas plastics and polymers still dominate, alternative materials have made significant progress in catching up. This is consistent with the growing number of use cases we discovered.

The difference among 3D printing in polymers and metals is not as great as you might imagine when it comes to basic use. More than one-third of respondents said they use plastics and metals equally, and even among those who said they use only one material, plastics only led by around 10% [36].

Naturally, obstacles still need to be overcome before some resources can be fully accessed. Nearly twice as many respondents as in 2019 said that it takes too long to develop the materials they need. A significantly higher percentage of respondents also mentioned that some materials are uncertified, unusable, or too expensive to utilise on a large scale.

1.4.3. POTENTIAL AREAS OF IMPROVEMENT

Besides all the advantages, still we are facing some problems about this printing technology. So researcher are utilizing their thoughts and times to overcome all the barriers like the usage of efficient print heads, the average 3D printer production speed is anticipated to improve by 88% by 2023, and as printer speed increases, volume capacities would also increase [27, 36]. And the other areas to improve as follows.

i) In today's era smart phones are very essential which account for around 35% of all consumer electronics sales, can be manufactured using 3D printing in the E&E sector. With greater growth anticipated in the near future

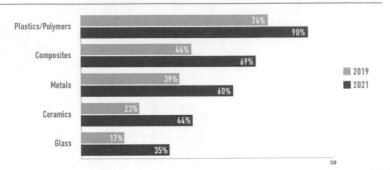

FIGURE 1.7   Growth Of Material In Recent Times.

Source: Reprinted with permission from [36]

due to improvements in materials and machinery, smartphone companies are gradually going beyond 3D printing's prototyping uses.

ii) In addition to enabling countless uses for 3D printing, new combinations of 3D printing materials and advancements to existing materials will also contribute to price reductions. By periodically impacting metal components with an ultrasonic tool while each layer is printed, metal parts can be strengthened. This releases the part's tensions, making it possible to print big things like aeroplane wings.

iii) Design skills must be improved because most 3D printers only produce objects with an efficiency of 94–96% percent due to atmospheric interference and printer hardware issues. If you want the exact shape, your design must have a tolerance of at least +− 0.5% [27].

REFERENCES

1. Abdulhameed, O., Al-Ahmari, A., Ameen, W., & Mian, S. H. (2019). Additive manufacturing: Challenges, trends, and applications. *Advances in Mechanical Engineering, 11*(2). https://doi.org/10.1177/1687814018822880
2. Ford, S., & Despeisse, M. (2016). Additive manufacturing and sustainability: An exploratory study of the advantages and challenges. *Journal of Cleaner Production, 137,* 1573–1587. https://doi.org/10.1016/j.jclepro.2016.04.150
3. Tofail, S. A., Koumoulos, E. P., Bandyopadhyay, A., Bose, S., O'Donoghue, L., & Charitidis, C. (2018). Additive manufacturing: Scientific and technological challenges, market uptake and opportunities. *Materials Today, 21*(1), 22–37. https://doi.org/10.1016/j.mattod.2017.07.001
4. Sathish, K., Kumar, S. S., Magal, R. T., Selvaraj, V., Narasimharaj, V., Karthikeyan, R., Sabarinathan, G., Tiwari, M., & Kassa, A. E. (2022). A comparative study on subtractive manufacturing and additive manufacturing. *Advances in Materials Science and Engineering, 2022,* 1–8. https://doi.org/10.1155/2022/6892641

5. Wasti, S., & Adhikari, S. (2020). Use of biomaterials for 3D printing by fused deposition modelling technique: A review. *Frontiers in Chemistry, 8*. https://doi.org/10.3389/fchem.2020.00315

6. Ingarao, G., & Priarone, P. C. (2020, August). A comparative assessment of energy demand and life cycle costs for additive- and subtractive-based manufacturing approaches. *Journal of Manufacturing Processes, 56*, 1219–1229. https://doi.org/10.1016/j.jmapro.2020.06.009

7. Dizon, J. R. C., Espera, A. H., Chen, Q., & Advincula, R. C. (2018). Mechanical characterization of 3D-printed polymers. *Additive Manufacturing, 20*, 44–67. https://doi.org/10.1016/j.addma.2017.12.002

8. Mikula, K., Skrzypczak, D., Izydorczyk, G., Warchoł, J., Moustakas, K., Chojnacka, K., & Witek-Krowiak, A. (2020). 3D printing filament as a second life of waste plastics—a review. *Environmental Science and Pollution Research, 28*(10), 12321–12333. https://doi.org/10.1007/s11356-020-10657-8

9. Rohde, S., Cantrell, J., Jerez, A., Kroese, C., Damiani, D., Gurnani, R., DiSandro, L., Anton, J., Young, A., Steinbach, D., & Ifju, P. (2017). Experimental characterization of the shear properties of 3D–printed ABS and polycarbonate parts. *Experimental Mechanics, 58*(6), 871–884. https://doi.org/10.1007/s11340-017-0343-6

10. Chua, C. K., Leong, K. F., & Lim, C. S. (2008). *Rapid prototyping*. World Scientific, p. 124.

11. Applied Engineering. (2022, May 19). Learn about the seven types of additive manufacturing, each with its own processes, methods of laye. *Applied Engineering*. www.appliedengineering.com/blog/2021/1/22/7-types-of-additive-manufacturing

12. Davis, R. (2021, August 18). What is 3D printing? Definition, technology and applications. *The Engineering Projects*. www.theengineeringprojects.com/2021/06/what-is-3d-printing-definition-technology-and-applications.html#

13. BCN3D Technologies. (2021, December 10). When was 3D printing invented? The history of 3D printing. *BCN3D Technologies*. Geraadpleegd op 1 mei 2022, van www.bcn3d.com/the-history-of-3d-printing-when-was-3d-printing-invented/

14. Hideo, K. (2014). Background of my invention of 3D printer and its spread. *Patent Magazine of Japan Patent Attorneys Association, 67*(13), 109–118.

15. Bhusnure, O. G. (2016). 3D printing & pharmaceutical manufacturing: Opportunities and challenges. *International Journal of Bioassays, 5.1*, 4723–4738.

16. Ferretti, P., Leon-Cardenas, C., Sali, M., Santi, G. M., Frizziero, L., Donnici, G., & Liverani, A. (2021). Application of TPU-sourced 3D printed FDM organs for improving the realism in surgical planning and training. *Proceedings of the international conference on industrial engineering and operations management* (pp. 6658–6669).

17. Wong, J. Y., & Pfahnl, A. C. (2014). 3D printing of surgical instruments for long-duration space missions. *Aviation, Space, and Environmental Medicine, 85*, 758–763. https://doi.org/10.3357/ASEM.3898.2014.

18. Rankin, T. M., Giovinco, N. A., Cucher, D. J., Watts, G., Hurwitz, B., & Armstrong, D. G. (2014). Three-dimensional printing surgical instruments: Are we there yet? *Journal of Surgical Research, 189*, 193–197. https://doi.org/10.1016/j.jss.2014.02.020.

19. Singh, S., Prakash, C., & Ramakrishna, S. (2019). 3D printing of polyether-ether-ketone for biomedical applications. *European Polymer Journal, 114*, 234–248. https://doi.org/10.1016/j.eurpolymj.2019.02.035.

20. Shotty, V., Dhagavan, K. R., & Ragothaman, A. (2016, April). Improved surgery planning using 3-D printing: A case study. *Indian Journal of Surgery, 78*(2), 100–104. https://doi.org/10.1007/s12262-015-1326-4

21. Ballard, D. H., Tappa, K., Boyer, C. J., Jammalamadaka, U., Hemmanur, K., Weisman, J. A., Alexander, J. S., Mills, D. K., & Woodard, P. K. (2019). Antibiotics in 3D-printed implants, instruments and materials: Benefits, challenges and future directions. *Journal of 3D Printing in Medicine, 3*, 83–93. https://doi.org/10.2217/3dp-2019-0007.

22. Tzounis, L., Bangeas, P. I., Exadaktylos, A., Petousis, M., & Vidakis, N. (2020). Three-dimensional printed polylactic acid (PLA) surgical retractors with sonochemically immobilized silver nanoparticles: The next generation of low-cost antimicrobial surgery equipment. *Nanomaterials, 10*. https://doi.org/10.3390/nano10050985.

23. Joshi, S. C., & Sheikh, A. A. (2015). 3D printing in aerospace and its long-term sustainability. *Virtual and Physical Prototyping, 10*(4), 175–185. https://doi.org/10.1080/174527 59.2015.1111519

24. Kalender, M., Bozkurt, Y., Ersoy, S., & Salman, S. (2020). Product development by additive manufacturing and 3D printer technology in aerospace industry. *Journal of Aeronautics and Space Technologies, 13*(1): 129–138.

25. Panneerselvam, P. (2018). Additive manufacturing in aerospace and defence sector: Strategy of India. *Journal of Defence Studies, 12*(1), 39–60.

26. 3D Printing Market. (2021). www.marketsandmarkets.com/Market-Reports/3d-printing-market-1276.html

27. International Journal of Engineering Research & Technology (IJERT). (2018). www.ijert.org ETEDM — 2018 Conference Proceedings.

28. Elakkad, A. S. (2019). 3D technology in the automotive industry. *International Journal of Engineering Research and Technology, V8*(11), 248–251. https://doi.org/10.17577/ijertv8is110122

29. FutureBridge. (2021, June 25). 3D printing – a technology that can print cars in future on a mass scale? *FutureBridge.* www.futurebridge.com/industry/perspectives-mobility/3d-printing-a-technology-that-can-print-cars-in-future-on-a-mass-scale/

30. International Journal of Engineering Research & Technology (IJERT). (2019, November). ISSN: 2278-0181 IJERTV8IS110122 (This work is licensed under a Creative Commons Attribution 4.0 International License), *8*(11). www.ijert.org

31. Nguyen, T. A., Gupta, R. K., & Behera, A. (2022). *Smart 3D nanoprinting: Fundamentals, materials, and applications* (1st ed.). CRC Press.

32. Guamán-Rivera, R., Martínez-Rocamora, A., García-Alvarado, R., Muñoz-Sanguinetti, C., González-Böhme, L. F., & Auat-Cheein, F. (2022). Recent developments and challenges of 3D-printed construction: A review of research fronts. *Buildings, 12*(2), 229. https://doi.org/10.3390/buildings12020229

33. Keita, E., Bessaies-Bey, H., Zuo, W., Belin, P., & Roussel, N. (2019). Weak bond strength between successive layers in extrusion-based additive manufacturing: Measurement and physical origin. *Cement and Concrete Research, 123*, 105787. https://doi.org/10.1016/j.cemconres.2019.105787

34. Ma, G., & Wang, L. (2018). A critical review of preparation design and workability measurement of concrete material for largescale 3D printing. *Frontiers of Structural and Civil Engineering, 12*, 382–400.

35. Zeif, Y. (2022, March 11). Four 3D printing trends to look for in 2022. *Fast Company.* www.fastcompany.com/90725018/four-3d-printing-trends-to-look-for-in-2022

36. Jabil. (2022). 3D Printing Trends: Six Major Developments. *Jabil.com.* www.jabil.com/blog/3d-printing-trends-show-positive-outlook.html

2 Basic Know-How About 3D Printing!

2.1. TYPES OF MAJOR 3D PRINTING TECHNOLOGY

In recent years, lots of 3D printing techniques are being used in several industries and home appliances.

2.1.1. STEREOLITHOGRAPHY (SLA)

Stereolithography (SLA) was first introduced method of an industrial 3D printing technology that can produce concept models, rapid prototypes, and complicated items with sophisticated geometries in much less time-close to a day or two [1]. SLA is known as the first 3D printing method, patented in 1986 by its inventor. Stereolithography parts can be made from various materials, with extremely high feature resolutions and high-quality surface finishes.

2.1.2. PROCESS OF SLA

Stereolithography is a single particle fabrication technology that produces shapes by selectively solidifying photosensitive polymers with UV light. There are two types of fundamental methods [2].

- Direct or laser writing
- Mask-based writing

These fundamental methods are mainly based on two basic approaches: free-surface and constrain-surface. The SL machine begins the 3D printing process by sketching the layers of the support structures first, then the part itself, using an ultraviolet laser aimed onto the interface of a molten thermoset resin. When one layer is placed on the resin surface, the build base descends and a coating bar moves over the platform to apply the subsequent layer of resin [2, 3].

2.1.3. MATERIALS FOR SLA

UV curable "resins" are thermoset polymers that are often used in SLA printing. They include a) Standard resin, b) Clear resin, c) Castable resin, d) Tough or Durable resin, e) High-temperature resin, and f) Dental resin.

Unlike previous generations of SL machines, today's machines have a variety of thermoplastic-like materials to pick from, including polypropylene, Acrylonitrile Butadiene Styrene (ABS), and glass-filled polycarbonate [1].

DOI: 10.1201/9781003349341-2

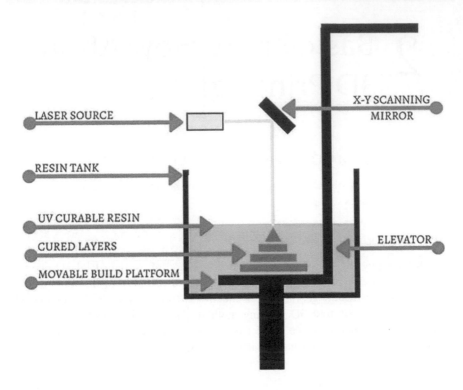

FIGURE 2.1   Stereolethography(SLA) Process.

Source: Reprinted with the permission from [1]

Polypropylene: A flexible, long-lasting resin that resembles hard polypropylene. It can endure abrasion and is ideal for fine detailing such as sharp corners, thin walls, and small holes.

Polypropylene/ABS blend: White, strong plastic that resembles a CNC-machined polypropylene/ABS combination. Snap fits, assembly, and demanding applications benefit from it.

ABS: Variations of ABS mimics encompass a clear, low-viscosity resin that can be finished clear; an opaque black plastic that blocks nearly all visible light, even in thin sections; a clear, colorless, water-proof plastic specific for lenses and flow-visualization models; and a micro-choice resin that lets in production of additives with especially tremendous capabilities and tight tolerances.

Polycarbonate: A ceramic-stuffed PC fabric that gives strength, stiffness, and temperature resistance, however may be brittle.

2.1.4.   BENEFITS OF SLA

Stereolithography is a type of 3D printing that has been in and around for a long time. The method aids medical device designers and other industrial innovators in visualizing and refining their notions in three dimensions. The following are some of the advantages of Stereolithography:

- SLA is capable of producing items with extremely high dimensional precision and fine detailing.
- They are suitable for visual prototypes because of the presence of a flawless surface finish.
- SLA materials, such as transparent, flexible, and cast able resins, are available.

2.1.5. LIMITATIONS

Like many engineering systems, SLA isn't a perfect fit, and it has several significant drawbacks when compared to alternative manufacturing methods. Many of the apparent difficulties of the procedure were connected to the storage and usage of the component elements. These are some of them:

- SLA pieces are fragile in general, making them unsuitable for practical prototypes.
- When SLA parts are exposed to sunlight, their mechanical qualities and visual appearance deteriorate over time.
- Support structures are always needed, and post-processing is required to eliminate the visual signs that the SLA component leaves behind.

2.1.6. APPLICATIONS

Speed, cost effectiveness, versatility, and accuracy are the advantages of stereolithography. These benefits make stereolithography a crucial procedure for making prototype that assists refines and prove designs in a variety of industries, including medical device design and 3D produced parts are used, amongst others, in automotive and wind tunnel testing.

2.2. DIGITAL LIGHT PROCESSING (DLP)

DLP System (Digital Light Processing) can be compared to stereolithography (SL) [4]. In that it's a 3D printing system that works with photopolymers. The important distinction is the light source. DLP makes use of an extratraditional light source [5], along with an arc lamp and a liquid crystal show panel, which is implemented to the complete floor of the vat of photopolymer resin in a single pass, typically making it quicker than SL.

2.2.1. PROCESS

The main parts of a DLP 3D printer are the digital light projection screen, DMD, vat (resin tank), build plate, and elevator for the build platform [5, 6].

- A direct digital projector serves as the DLP 3D printer's light source.
- The digital light projector uses the DMD (Digital Micromirror Device), a part made up of a lot of micromirrors, to guide the light source it projects.

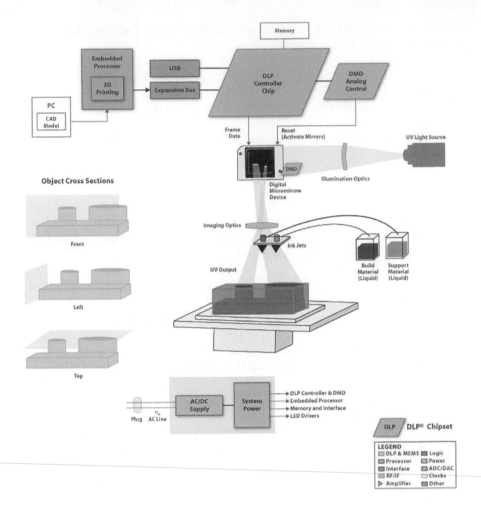

FIGURE 2.2 Digital Light Processing 3D Printing.

Source: Reprinted with permission from [5]

- The next item on the line is the vat, which is simply a resin container. However, the bottom of the vat needs to be transparent so that when the resin may be cured by the light from the digital light projector.
- The surface that printed objects stick to throughout the printing process is actually referred to as the build plate. The self-explanatory element known as the z-axis [7] is used to gradually increase the construction platform it during printing process.

2.2.2. MATERIAL

The procedure and availability of materials varies in different devices, more than it varies in different technologies, with accuracy and precision. Liquid photopolymers mostly in form of resin are used in DLP 3D printers. There are numerous resins available, within a wide range of prices.

Standard Resin: The most common liquid resin for additive manufacturing processes is standard resin. Its excellent quality with amazing detail and inexpensive pricing are the major reasons for its widespread use [7].

Clear Resin: It belongs to the same group as Standard Resin, but due to its distinct hue, it warrants its own category. Clear hues, like clear red, clear greenish, clear blues, or just the clear hue itself, are available in several resins. Furthermore, these resins are moisture-resistant, albeit the model's clarity may deteriorate over time if it is exposed to UV rays.

Tough Resin: When it comes to creating durable, useful resin 3D prints that can withstand stress and strain, Tough Resin is an excellent choice. It's also known as ABS-like because of its mechanical qualities, which are similar to those of ABS [7].

Because of their exceptional flexibility and ease of post-processing with water rather than alcohol or other solutions, flexible resin and water washable resin are also employed in this technique.

2.2.3. Pros of DLP Method

Digital light processing (DLP) is advantageous since it allows high-power intensities and the simultaneous illumination of the whole printing platform. Engineers nowadays employ this technology for the following reasons:

- DLP printers feature superior surface qualities and are quite precise as comparing to other kinds of printers. Those surfaces can be treated to a high sheen and are visually appealing [7].
- Though they are occasionally comparable, DLP prints require less time than SLA prints due to their uniform layer projection.
- DLP parts are isotropic, which means that they are equally strong in all the three planes (X,Y,Z), whereas FDM parts are not. DLP parts are also completely waterproof and largely unresponsive to anything other than light.
- Sintered metal parts can be printed using specialised DLP printers, which eliminates many of the material's drawbacks (for an increased cost).

2.2.4. Limitations of DLP Method

SLA 3D printers are equivalent or faster for printing tiny or medium-size single pieces, while DLP 3D printers are faster when manufacturing huge, completely dense prints, or structures with several parts that cover up much of the platform, according to the study. But this technology also has some disadvantages which are as follows:

- Due to their brittle nature, DLP prints are less effective for functional prototypes and are better suited to visual prototypes.

- Because sunlight degrades the aesthetic and mechanical qualities of resin parts, DLP printed parts are unsuitable for outdoor use.
- DLP printers generate voxelated pieces that must be sanded and smoothed to match the surface quality of SLA parts [6, 7].

2.2.5. APPLICATIONS

DLP 3D printing technique is quite similar to SLA in terms of operation and results. DLP printers have a wide range of applications since their introduction in the late 1980s, and they continue to do so today. These days, the use of technology is quickly increasing because of the following factors:

- DLP is well suited to the dentistry applications, and it can print surgical guides, crowns and bridges, and orthodontic devices more quickly.
- When a surgical guide, or even aligner is inserted into a patient's mouth, the patient's saliva comes in touch with the foreign object, making it necessary to use only such materials that do not harm the patient. As a result, material scientists have produced bio-compatible resins that have no harmful effects on the patient throughout the process. In DLP 3D printing, this sort of resin is employed for fast results [7].
- Medical prototypes and even precise medical gadgets are frequently recreated using DLP 3D printing technology. Bones, Genetic study samples, tooth models, muscle models, and other models are used for instance.
- By generating wax components of the design immediately, a time-consuming first step of creating a mould can be avoided, saving time and money.

2.3. FUSED DEPOSITION MODELLING (FDM)

The additive process includes the Fused Deposition Modelling (FDM) method [8]. It's one of several additive processes that include Stereo lithography (SL), Selective Laser Sintering (SLS), and others. FDM, as the name implies it refers to the process of layering fused materials to create parts [8]. Modelling, prototyping, and production applications all leverage this Rapid Prototyping approach. S. Scott Crump created FDM in the late 1980s, and Stratasys commercialized it around 1990.

2.3.1. PROCESS

This technique involves moving the heated printer extrusion head to move the filament from a large bobbin and deposit it on the developing item. The print head is moved by the computer-preset programmes to specify the printed shape. Typically, filament comes in two common diameters: 1.75 mm and a few 2.85 mm. Although 2.85 mm filament is frequently and incorrectly called "3 mm," it shouldn't be mistaken with the much smaller filament size, which effectively addresses 3 mm in diameter [8]. The print head typically moves in two directions to drop one horizontal layer or plane at such a time, then the work or print neck is adjusted very little vertically to create a second layer.

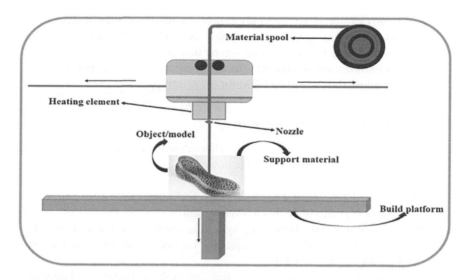

FIGURE 2.3 Fused Deposition Modelling.

2.3.2. MATERIAL

The majority of the current FDM machines use filament-based thermoplastics for ejection and deposition. Most of the time Polylactic Acid (PLA), Acrylonitrile Butadiene styrene (ABS), Polycarbonate (PC) are used in this process. But nowadays Thermoplastic Polyurethane (TPU), Polyethylene Terephthalate (PET) or even Metals are also used in this technology. Water-soluble wax or brittle thermoplastics, such as polyphenylsulfone (PPSF), are commonly used as support materials. Several materials such as Polypropylene (PP) and Polyethylene (PE) have biocompatibility and MRI transparency, which are used in medical applications.

The previously mentioned materials are used in two parts:

- Build material: This substance is used to create the final product.
- Support material: This material is utilized to build the primary part's support structures. If the base material is dissolvable, it is referred to as dissolvable material.

2.3.3. ADVANTAGES OF FDM TECHNIQUE

FDM can construct prototyped items and parts rapidly and economically using a thermoforming method. Designers and engineers typically use it as a non-laser-based approach to build and perfect their concepts before mass manufacturing [9]. They choose FDM process because of the following:

- With this type of 3D printing, objects are automatically resized down to fit inside a production space while maintaining precision. Micro samples can be created for use in demonstrations or as scaled-down samples in sales.

- FDM parts can be ready in a two minutes or hours, making it the fastest method of 3D printing among the options available. CAD drawings can be turned into final goods in a single step by this method.
- Waste minimization, economic effectiveness, and environmental friendliness are all reasons to utilize FDM for 3D printing.

2.3.4. LIMITATIONS OF FDM TECHNIQUE

According to experiments, as compared to alternative manufacturing methods, this popular technology has numerous notable downsides.

- These printed items have uneven surfaces. Techniques used after curing are one approach to eliminate this uneven terrain.
- When using an FDM 3D printing system to create large flat surfaces, the warping fault occurs the most frequently. Layers can stretch in any direction because of their flexibility.
- FDM 3D printers are based on the idea that melting filament and depositing it layer by layer on top of one another will create a three-dimensional structure. Now, when carrying out this process, filament could get stuck in the nozzle and cause nozzle blockage.

2.3.5. APPLICATIONS

FDM printed products are frequently used for a huge number of utilizations across industries from prototypes model to final parts.

- Scaffold development: Scaffold (a porous structure) works as a synthetic [9] extracellular matrix in tissue engineering, supporting cell adhesion, proliferation, and differentiation.
- Aerospace and automotive applications: FDM in the aircraft industry has became popular since fire-retardant and flame-retardant polymeric models have been made into use.
- Drug delivery systems: FDM is the cheapest among these processes and polymeric drug delivery vehicles (DDV) and can be efficiently generated from it. Water-soluble and hydrophilic polymers are best suited as DDV [9].

2.4. SELECTIVE LASER SINTERING (SLS)

Dr. Carl Deckard and his academic mentor, Dr. Joe Beaman, developed and patented Selective laser sintering (SLS) at the University of Texas at Austin in the mid-1980s [10]. SLS (like with the other AM processes described) is a fairly new concept that has primarily been utilised for fast prototyping and low-volume component part manufacture.

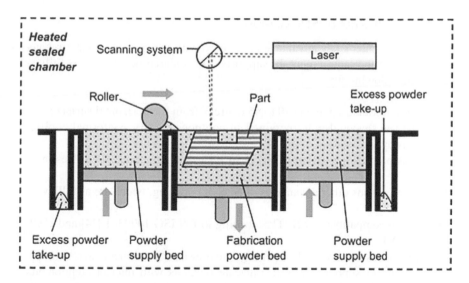

FIGURE 2.4   Selective Laser Sintering (SLS) Process.

Source: Reprinted with permission from [11]

2.4.1.   PROCESS

Selective laser sintering (SLS) is one type of additive manufacturing (AM) process that fuses powdered materials (usually nylon or polyamide) while autonomously targeting the laser at locations in space specified by a three-dimensional model. The procedure typically includes the following steps.

- A laser beams down on the base, drawing a cross-section of the item onto the powder under the direction of a system that specifies which object to "print."
- The laser beam warms the powder to a temperature that is either a little below its boiling temperature (sintering) or above it (melting), which causes the powder's components to fuse with each other to form a solid.

2.4.2.   MATERIAL FOR SLS

Because of their perfect sintering behaviour as a semi-crystalline thermoplastic, polyamides are the most often utilised SLS materials, resulting in parts with desirable mechanical qualities [12]. Polymers such as polyamides (PA), polystyrenes (PS), thermoplastic elastomers (TPE), and polyaryletherketones (PAEK) are among the commercially accessible materials utilised in SLS and are available in powder form [13]. Due to its high toughness, thermal stability, and flame resistance, polycarbonate (PC) is a useful interest for SLS; however, such amorphous polymers processed by SLS tends to result in parts with diminished mechanical properties and dimensional accuracy, and are hence restricted to cases where these are not crucial [12].

2.4.3. Advantages of SLS Method

The SLS 3D printer's key advantage is the ease with which it can create in batches and the lack of the requirement for supports. The sintered powder bed is fully self-supporting, allowing for:

- Wide overhanging angles (0 to 45 degrees from the horizontal surface).
- Intricate shapes deeply ingrained in components, like conformal cooling pathways.
- Mass-producing a variety of pieces in 3D arrays, a method called "Nesting".

There are numerous other advantages of SLS 3D printing, which are as follows:

- It's bio compatible at 121 °C according to EN ISO 10993–1 [14] and USP/level VI
- Parts are extremely rigid and strong, and they have good chemical resistance.
- Because of support removal, intricate portions with interior components can be built without trapping material or altering the surface.
- Because of their dependable mechanical properties, parts can frequently replace conventional injection-moulded plastics.

2.4.4. Disadvantages of SLS Method

The biggest downside of the SLS 3D printer is the limitation of various raw material alternatives, as well as the cleanliness procedure involved after printing. There are several other limitations, which include:

- It is a very much expensive. Machines can cost up to $250,000, with materials ranging from $50 to $60 per kilogram. Furthermore, the machineries need the use of competent operators [15].
- It takes almost 50% of the print time to cool down, extending approximately to 12 hrs. As a result, the production time is extended. Furthermore, because the pieces have a rough surface, post-processing is required.

2.4.5. Applications

The manufacture and development of functional components are both possible with selective laser sintering. This is due to the fact that SLS is known for its high precision and ability to print complicated shapes [15]. Early stages of the design cycle, it's most commonly used in prototype parts like investment casting designs, automobile components, and wind tunnel models. SLS has also been utilised to make end-use components for aircraft industry, military, medical, pharmaceuticals, and electronics gear in limited-run fabrication.

2.5. ELECTRON BEAM MELTING (EBM)

For metal items, electron-beam melting (EBM) is a sort of additive manufacturing, or 3D printing. The crude material (metal powder or wire) is heated by an electron

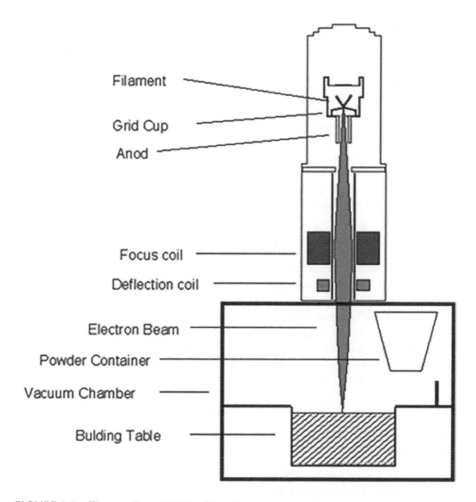

FIGURE 2.5 Electron Beam Melting Procedure.

Source: Reprinted with permission from [10]

beam and fused together under a vacuum. This method differs from selective laser sintering in raw material which is entirely melted before it is fused [16].

2.5.1. PROCESS

In the very first step, we need the 3D model of our desired part. You can create a model with CAD software, 3D scan it, or download a model of your preference. After that, the model is then transferred to a slicing software, which is known as a slicer. This software cuts the model into layers based on the physical layers of deposited material. The slicer will then transfer all of this data to the 3D printer, allowing it to begin the production process. The machine's tank can be loaded with metal powder. It will be placed in tiny layers that will be warmed before the electron beam fuses them. This step, adds more support to the cantilever portions of the 3D-printed item. The printer then repeats these operations as many times as necessary until the full portion is obtained.

2.5.2. MATERIALS

The materials required must be conductive because the procedure is built on the basis of electrical charges. There will be no interaction between the electron beam and the powder if this is not done. As a result, using an electron beam to make polymer or ceramic pieces is completely impossible, and therefore only metals can be employed. Commercially pure Titanium, Ti-6Al-4V [17], CoCr, Inconel 718 [18], and Inconel 625 [19] are currently available for EBM.

2.5.3. ELECTRON BEAM OR LASER?

The answer is mostly dependent on the applications you're interested in, as each technique has its own set of advantages and disadvantages.

- The electron beam may split and heat the powder at multiple locations at the same time, significantly speeding up production [20]. A laser, on the other hand, must scan the entire surface point by point.
- The electron beam is just a little broader than the laser beam at the powder level, which affects precision.

So, for our desired need, we must indicate whether we want to use a beam or a laser.

2.5.4. APPLICATIONS

- EBM technology is mostly applied in the fields of aeronautics and medicine, especially in implant design.
- Titanium alloys are particularly intriguing due to their biocompatibility and mechanical qualities, as well as their ability to produce material that is lightweight and has great strength.
- The method is frequently utilized in the design of turbine blades and engine elements.

2.5.5. PROS OF EBM PROCESS

Like any other additive manufacturing technology, EBM has benefits that make it suitable for specific applications. Most metal 3D printing techniques use a laser as their major heat source, but this process employs a beam multiple times stronger. The higher beam power leads to faster printing speeds. EBM can create high-quality pieces of metal that are comparable to those created using traditional methods such as casting. The components appear to have strong mechanical characteristics as a result of the pre-heating procedure and the high temperatures reached during printing, but they also often have a greater density [21]. EBM technologies also have other advantages, which are as follows:

- EBM produces very little waste since most of the wasted powder can be reused, so that is especially advantageous given the high cost of the ingredients used in EBM.
- Tooling and setup of this technique are less expensive and user friendly.
- The mixing of the powder is done in a vacuum environment so oxidation is reduced.

2.5.6. Cons of EBM Process

EBM is not restricted to certain flaws and limits. When compared to other 3D-printed components, EBM items have a lesser degree of precision. Other disadvantages of EBM are as follows:

- To get a better surface quality, manufactured pieces require substantial post-processing.
- The materials employed in the EBM method are likewise limited; this is partially due to the fact that the technique requires high-quality, pricey materials that must also undergo extensive testing before usage.

2.6. BINDER JETTING AND MATERIAL JETTING

Whereas there are numerous additive manufacturing technologies with a wide range of applications, some can share some common features. 3D printing methods, for comparison purposes, frequently overlap in areas, such as with the materials utilized, the heat source used, the print head, and so on. Material jetting and binder jetting, both of which project some form of substance onto a plate, are two examples of this. A binder is "jetted" into a bed of powder in Binder Jetting, causing the granules to fuse together, whereas droplets of photosensitive resin are applied and subsequently solidified using UV light in Material Jetting [22].

2.6.1. The Principle Behind Material Jetting

Material jetting acts on the very same concept as two-dimensional printing. Multiple print heads release material droplets on a printing surface, similar to how a conventional 2D printer transfers ink to a piece of paper. The substance utilized in our situation is photosensitive resin or photopolymer, which is solidified by passing UV light across the print surface [22]. The material is deposited within a single encounter. After that, the print platform is lowered and the process starts all over again. Furthermore, it has a quick printing speed. Material Jetting, like SLA and DLP, involves the use of support systems made of a soluble material that the user eliminates when the printing is finished.

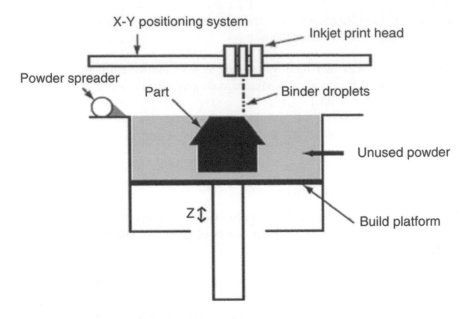

FIGURE 2.6   Binder Jetting Process.

Source: Reprinted with permission from [23]

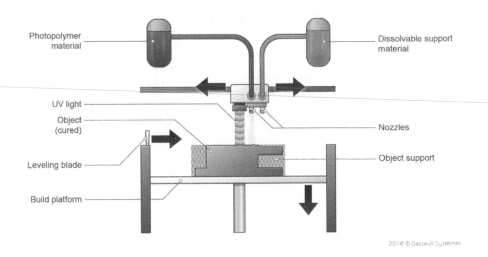

FIGURE 2.7   Material Jetting Process.

Source: Reprinted with permission from [24]

2.6.2.   What Is Binder Jetting and How Does It Work?

Binder jetting is a technique that employs the use of binders and powders. In concrete terms, the machine applies a binder to a coating of powder, which works as an adhesive and allows the particles to be bound together. The technique is continued until the final component is achieved, layer by layer [25]. Binder jetting, like laser sintering, doesn't need any kind of print backing because the powder serves as the support. After the printing is completed, the powder around the component must be removed, followed by the removal of all leftover powder and cleanup of the part.

2.6.3. MATERIALS FOR JETTING

Because of their good mechanical properties, high corrosion resistance, and high chemical resistance, full color sandstone, silica sand, stainless steel (bronze infiltrated or sintered), sintered inconel alloy, and sintered tungsten carbide are frequently used as powders in this Binder jetting technique. However, these materials have several disadvantages, such as brittleness and porosity inside the substance.

Whereas Material Jetting employs thermoset photopolymer resins with features and restrictions comparable to those used in SLA [26]. Photopolymers (in liquid state) and casting waxes are the major most prevalent materials utilised in Material Jetting nowadays.

2.6.4. APPLICATIONS

MJ's multi-material printing capability allows for the construction of precise visual and tactile prototypes. Material Jetting's multi-material capabilities may be used by designers to create very accurate, full-color visual prototypes [27]. Material Jetting is becoming more popular in the medical field for creating effective anatomical prototypes for educational reasons as well as pre-surgical selection and training. Material jetting may also be used to print electrical components in three dimensions. In today's market, Binder jetting is also often used in areas like the aircraft industry and heavy manufacturing, where high thermal stability and wear-resistant components are required. Binder Jetting might potentially be used in the food business, which is an uncommon but intriguing use.

2.6.5. ADVANTAGES AND DISADVANTAGES OF BINDER JETTING AND MATERIAL JETTING

The main advantage of this technique is it has the capacity to create multi-material and multi-color components, which is the fundamental benefit of the Material Jetting technique. In most cases, the technique is shorter than others. The two-material method allows for a variety of binder–powder combinations as well as different mechanical properties [28].

Binder jetting, on the other hand, is not ideal for structural parts due to the binder material [28], and post-processing might add significant time to the overall technique. Low waste due to precise jetting and material on-demand dropping technology, as opposed to powder bed fusion, which sinters inside the power chamber, makes it very much user friendly. But mechanical properties of photosensitive material used in PolyJet degrade quickly any time and produce relatively brittle parts, making it difficult to use in a production load-bearing part [29].

2.7. DROP ON DEMAND

It is a brand-new 3D printing technique that makes use of two inkjet printers. The construct material, which is often a wax-like substance, is deposited from the one inkjet. For dissolved support material, utilise the second inkjet. Like other common 3D printing technologies, DOD printers use a preset path to jet material in a point-by-point deposition, building up the object's cross section layer by layer [30].

Another feature of DOD printers is indeed a fly-cutter, which basically takes the construct area after each layer is finished to guarantee a completely flat surface before beginning the next layer. For lost-wax casting, investment casting, and other mould-making applications, DOD printers are typically employed to produce patterns [31].

2.7.1. PRINCIPLES OF DROP ON DEMAND PRINTING

In this process, the ink is heated until entirely evaporated. As a result of this expansion of the ink droplets, ink drops will appear on the printing plate. Keep in mind that each 3D printing nozzle is independently managed, so if one nozzle malfunctions, it won't affect the other nozzles [32].

Because it prints over for a surface of an object, drop-on-demand printers are recommended in this situation. As a result, the item will no longer need labeling. The benefits of drop-on-demand 3D printing are as follows:

i)   Remember that Drop on Demand 3D printers have big nozzles. Large font sizes can be printed with the large-diameter inkjets.

ii) In inkjet printing, the printed item will be clean, clear, and legible.

iii) Knowing that inkjet printers function effectively with specific connectivity is a wonderful thing. Therefore, customers can print various colours on cardboard and plastic materials using drop-on-demand inkjet printers. Additionally, it ensures printing of a high standard.

iv) Inkjet printers can be utilized over an item, as we mentioned previously. As a result, the expense of creating labels is removed [31, 32].

2.8. THE DESKTOP 3D PRINTER

Why would a typical customer need a 3D printer? Many of the technology's early adopters were amateurs who wanted to test it out for fun. By participating in open-source forums, they have advanced printer design. Toys, tool holders and organisers, science fiction character figurines, role-playing game props, abstract mathematical sculptures, and add-on or upgrade parts for 3D printers themselves have been the stereotypical uses of printers among the hobbyist community [33].

New buyers, nevertheless, are also beginning to show up with important applications. A 3D printer may be used to create a physical prototype that is easier to modify and evolve with clients than a conventional foam-core or clay one, making it ideal for product designers, Hollywood special effects artists, and other professionals who frequently need one-off models [34].

2.8.1. Types of Filament-Based Consumer Printers

As said before, this book concentrates on printers that melt filament from spools or cartridges of material, mostly either 1.75 mm or 3 mm in diameter and build the melted material up layer by layer to produce objects due to their simplicity and accessibility. However, resin printers for home use are now available, including models from Formlabs and B9Creations. Using one of these is very different from using a printer that uses filament. Currently, the majority of desktop 3D printers using filament are Cartesian. Cartesian printers are named when the Cartesian [35] reference system uses X, Y, and Z coordinates to plot points. These machines feature a framing that is roughly rectangular in shape, or at least is constructed from sections that are at right angles. To produce a 3D print, various printer components move along each of the axes. Cartesian printers are usually diagonally supported even though they travel along right-angled x-, y-, and z-axes.

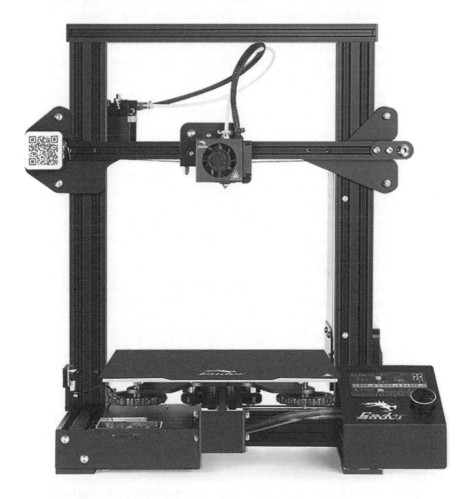

FIGURE 2.8 Filament 3D Printer.

Apart from that, different kinds of non-Cartesian 3D printers also exist. The Deltabots [36] are a fascinating non-Cartesian 3D printer family tree branch. They have roots in pick-and-place industrial robots, which are used in manufacturing to precisely put a part on an assembly or within a package. The following subsections contrast Deltabot's and Cartesian's design approaches [37].

René Descartes created the Cartesian coordinate system, which is where the name "Cartesian printer" comes from. They not only work according to an X and Y coordinate system, but they also move the print head linearly along the x- and y-axes. This type of printer prints horizontally, vertically, and from front to back. Simply explained, it moves one axis at a time to move from point A to point B. The majority of Cartesian printers have a square, movable print bed. In some instances, the print bed advances towards the nozzle rather than the other way around [38].

While Delta 3D printers similarly employ the Cartesian coordinate system, they do not deposit the filament using controlled linear motion. Instead, they employ three parallelogram-shaped arms. Along with moving between X and Y points, they also change the angles of these parallelograms. A fixed platform supports the hanging arms of a Delta 3D printer. The platform contains every mechanical component that propels the arms' movement. The print head of the Delta 3D printer is much more lightweight as a result than it would be if it also had to house motors for movement. Reduced inertia results from this decreased weight. The output head can react quickly while maintaining precision by decreasing inertia, especially toward the end of a movement [37, 38].

2.8.2.   3D Printer Design Considerations

To help you understand what we will be discussing later on, we will now briefly describe some of the printer's components that will be covered in the remaining chapters of the book. It's been compared to a computerised hot glue gun; a filament-based 3D printer literally melts polymer filament and deposits it in layers. This has a number of important components that necessitated numerous design trade-offs amongst the available solutions.

- **Filament:** Filament is often supplied in proprietary cartridges or on spools (for open-source printers). Printers must include a method for removing filament from the spool, such as by spinning the filament on a spindle or a tray similar to a lazy Susan. In some instances, this is affixed to or a part of the frame; in other instances, the filament is supported by a lazy Susan next to the printer. Filament is commonly available in 1.75mm and 2.85mm diameter, but most of the time 2.85mm is considered as a 3mm [39].
- **Frame:** For such prints to be precisely built up, the printer's frame must be rigid and strong. It is unlikely that you will be able to get an appropriate print if the frame is sloshy. Typically, frames are constructed from a variety of extruded aluminium rail kinds. The remainder of the frame is often composed of laser-cut or 3D-printed joints.
- **Build Platform:** To construct the print, every 3D printer needs a level surface. Common names for this area include the build platform, platform, and bed. To enable printing of materials that need to be kept warm throughout

construction, certain printers incorporate heated platforms. Others only require some form of tape to be placed over a glass or other plate in order to make sure the first layer of the print will adhere. There are times when a Kapton high-temperature tape is used to wrap the printer bed.

- **Extruder:** The component of the printer that melts and pushes the filament is called the extruder. There are various parts to the extruder. The first is an extruder drive system, which consists of a motor and a device that forces filament into the hot end. The hot end itself is made up of a heater, a nozzle, and a sensor to measure how hot the bed is (a thermistor). Figure 2.1 depicts a dual-extruder printer with two extruders that can print things from two different materials.
- **Bowden and Direct-Drive Extruders**: There are numerous extruder designs, which can be divided into two main groups. A motor and driving gear are located right next to every hot end of a direct-drive extruder, like the one on the machine in Figure 2.1. A guide tube separates the drive gear from the hot end of a Bowden extruder (like the one on the machine in Figure 2.2). Using a Bowden extruder is beneficial because it takes the bulky motor away from the nozzle, making the moving portion of the extruder much lighter. This might make it possible to print more quickly, but at the expense of a more complicated extrusion system.
- **Retraction:** There are voids and filled areas when the printer is building a layer. Some of the holes can be created by zipping the nozzle all the way around them. However, there are situations when it's vital to halt extrusion over a layer gap. In that situation, the filament must be retracted by the extruder. To make this work, the drive gears must be able to both pull and push filament ahead. A user can specify how much to retract during a print when the model is being divided into layers. Print quality is strongly impacted by retractions.
- **Nozzle:** One of the really important and delicate components of the printer is the nozzle. The tiny, easily clogged openings. What materials your printer can safely melt and print, in part, depend on the type of nozzle the printer has. Polycarbonate, nylon, and other higher-temperature plastics are manageable by high-quality nozzles. The thermal break (barrel), the heater block, and, in many designs, a heat sink to cool the top of the thermal break are all components of the hot end, which also contains the nozzle. Occasionally, ceramics or high-temperature polymers' insulating capabilities are used in certain designs.
- **Moving Parts:** 3D printers need to move some combination of the extruder and the build platform to be able to create objects. They achieve this with a combination of stepper motors attached to drive screws or cables, belts, or other systems attached to pulleys. Stepper motor is nothing but a brushless DC electric motor that separates a full rotation into a number of equal steps, also known as a step motor or stepping motor. As long as the motor is appropriately scaled for the application in terms of torque and speed, the location of the motor can be instructed to travel and stay at a particular step without any position sensor for feedback (an open-loop controller).

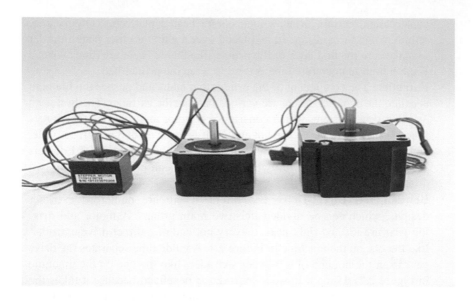

FIGURE 2.9 Stepper Motor.

- **Control System:** Microprocessors that work with Arduino are commonly used to operate open-source printers. The control systems are quite straightforward. The majority of 3D printers for consumers are open loop systems. This means that even while the microprocessor transmits a command, it has no way of knowing whether it was correctly carried out or not. The general open loop technique has a few exceptions. A thermistor temperature sensor is present on every heated component of a printer, including heated beds and extruders. Without this, it would be difficult to control temperature. The majority of printers also include end stops. These enable the machines to "home" to a known spot, after which they may measure their distance from that location by counting the number of steps they have taken in each direction [39–41].

2.9. OPEN SOURCE

Any programme whose raw data or source code is freely available for usage or customization by users or other programmers is referred to as being "open source." Open-source software, as opposed to proprietary software, is computer software that is created through a public, open cooperation and made publicly available for free [42].

But why would someone publish the source code of their programme or hardware? Most individuals who create open-source code do it by way of their employment. The majority of network administrators and programmers who require dependable, affordable Web server software and think it's better to combine efforts than go it alone originally wrote Apache, and they continue to do so today [43]. And also, it is a really effective technique for technological groups to explore and discover good ways through many different potential growth paths. Open source frequently connotes total charity combined with a desire for praise from the community. In other words, there are several underlying reasons for writing open-source software aside from altruism or the sheer delight of creativity, and some of those motivations are just as egotistical as writing proprietary code because you get paid, which can be (and frequently is) a great motivation for writing open-source software.

2.9.1. HISTORY OF OPEN SOURCE

The term "free software" (with the emphasis on freedom) first appeared in the early 1980s. Open source is a considerably more recent concept, dating from the late 1990s.

However, certain applications paved the way before free and open-source software (FOSS) existed as such. In truth, up until the late 1960s, the majority of software was free and open source (FOSS) and could be easily shared among those who maintained computers [44]. The idea of open source eventually fell out of favour during the 1970s to 1980s as software became more commercially viable. But collaborative software development is still common among academics. Examples include Richard Stallman in 1983 with the GNU operating system and Donald Knuth in 1979 with the TeX typesetting system [45]. "The Cathedral and the Bazaar," a reflective examination of the hacker culture and free-software tenets, was released in 1997 [46] by Eric Raymond. Early in 1998, the paper attracted a lot of attention, which inspired Netscape Communications Corporation to make its well-known Netscape Communicator Internet suite available as free software. Later, this source code served as the foundation for KompoZer, Mozilla Firefox, SeaMonkey, and Thunderbird.

2.9.2. GNU LICENSE

The four rights to run, study, share, and alter the software are guaranteed under the GNU General Public License (GNU GPL), a group of commonly used free software licences. Richard Stallman, the man behind the Free Software Foundation (FSF), wrote the licence as the first copyleft for general use specifically for the GNU Project.

It gives users open-source permissions such as:

- It's able to freely access and use the programme.
- It's able to modify the software as desired.
- We will be capable of sharing copies of the software.
- Able to alter and share copies of new software versions. [47]

2.9.3. HOW DO WE USE GPLv3?

Users must get a copyright disclaimer from whatever superseding entity, such as a workplace or school, in order to release software under GPL licences. Once the disclaimer is in place, each file should be properly copied with notices that specify which versions users are permitted to use [48].

Add licence disclosures and a declaration of permission towards each file, and then include a COPYING file with the entire text of the GNU GPL terms and conditions. It is not required to display the copyright notice at launch.

2.9.4. PROS AND CONS OF GNU LICENSE

The fact that it is simple to update code, documentation, or hardware blueprints, depending on the situation, is truly the root of both the benefits and drawbacks of open-source software. This implies that a flaw or error can be swiftly fixed by the community [49]. However, if you plan to use an open-source platform, you must carefully consider how to coordinate system updates and configuration management with the status of the open-source software. Large, widely used programmes may eventually, in one way or another, employ one or two persons to oversee the overall programme and carry out controlled version releases. The completeness and versioning of open-source documentation are relevant problems. There are several programmes with excellent documentation that matches or even exceeds the requirements of commercial documentation. Others might provide extensive information on specific capabilities that is suitable for advanced users but lacks "big picture" documentation that would be helpful for a first-time user who is attempting to figure out where to begin. The documentation can be outdated compared to the software [50].

2.9.5. MEETING THE OPEN-SOURCE 3D PRINTING COMMUNITY

Since it lowers software costs and broadens distribution, open-source software has been a huge success. As a component of the open-source movement, open-source hardware has recently gained popularity due to its potential to further reduce the cost of all types of manufactured items and alter the manufacturing supply chain. A single-board open-source microcontroller named Arduino is a hardware part of a 3D printer. Since the Arduino microcontroller is an example of open-source hardware, any skilled engineer can alter, expand, or even create their own board based on the open-licensed design. The board was initially created by Massimo Banzi and his colleagues to assist his students with their projects; however, it quickly gained popularity because of its open design, affordable pricing, and incredibly simple programming style. Today, Arduino has developed into a series of microcontrollers that offers solutions for many types of applications, from microcontroller to microcomputer, ranging from basic command to multi-media. AutoCAD, an open-source computer-aided design (CAD) programme, is used to create the schematics for the open-source optics hardware. A command-based CAD toolbox called AutoCAD enables users to build three-dimensional structures by writing code or scripts in the dialogue box. The user should be knowledgeable with the syntax and commands to construct and change 3D objects, but little programming experience is necessary. Design in AutoCAD begins with basic, widely used 3D shapes known as primitive shapes, such as spheres, boxes, cylinders, and polygons. From these fundamental shapes, complex 3D objects are derived using a variety of commands, including combination, extrusion, Boolean expression, and many others. The main benefit of AutoCAD is its parametric modelling, which parameterizes an object's geometrical specification and transmits relational characteristics to its derivatives. This enables straightforward customization of the produced objects by just changing the user-defined variables [51].

REFERENCES

1. Stereolithography (SL) Concept Models, Cosmetic & Prototypes. (2022). *www.protolabs.co.uk*. Geraadpleegd op 5 mei 2022, van www.protolabs.co.uk/services/3d-printing/stereolithography/
2. Manufactur3D. (2022, February 27). How stereolithography Or SLA 3D printing works? *Manufactur3D*. https://manufactur3dmag.com/stereolithography-sla-3d-printing-works/
3. THINK3D. (2022, April 16). *Stereolithography (SLA) 3D printing services in India*. Geraadpleegd op 1 mei 2022, van www.think3d.in/services/3d-printing/stereolithography/
4. www.thomasnet.com/articles/custom-manufacturing-fabricating/digital-light-processing-dlp-3d-printing/
5. Ali, M. H., Batai, S., & Sarbassov, D. (2019a). 3D printing: A critical review of current development and future prospects. *Rapid Prototyping Journal, 25*(6), 1108–1126. https://doi.org/10.1108/rpj-11-2018-0293
6. Digital Light Processing 3D Printing Explained. (2019, October 17). *Wevolver*. Geraadpleegd op 1 mei 2022, van www.wevolver.com/article/digital.light.processing.3d.printing.explained
7. Sathies, T., Senthil, P., & Anoop, M. S. (2020). A review on advancements in applications of fused deposition modelling process. *Rapid Prototyping Journal, 26*(4), 669–687. https://doi.org/10.1108/RPJ-08–2018–0199
8. Deckard, C. (1989, September 5). Method and apparatus for producing parts by selective sintering. U.S. Patent 4,863,538, filed October 17, 1986.
9. Kloos, S., Dechet, M. A., Peukert, W., & Schmidt, J. (2018, July). Production of spherical semi-crystalline polycarbonate microparticles for Additive Manufacturing by liquid-liquid phase separation. *Powder Technology, 335*, 275–284. https://doi.org/10.1016/j.powtec.2018.05.005.
10. High-end plastic materials for additive manufacturing. Retrieved February 19, 2019, from www.eos.info.
11. EBM-Built Materials-Arcam AB (2013, January 24) Retrieved April 26, 2017, from Arcam.com.
12. International Organization for Standardization (ISO). (2009). Biological evaluation of medical devices—part 1: Evaluation and testing within a risk management process (ISO 10993-1:2009). OCLC 839985896.
13. 3DSourced. (2021, June 26). Selective laser sintering: Everything you need to know about SLS 3D printing. *3DSourced*. Geraadpleegd op 1 mei 2022, van www.3dsourced.com/guides/selective-laser-sintering-sls/
14. Leong, K. F., Liu, D., & Chua, C. K. (2014). Tissue engineering applications of additive manufacturing. *Comprehensive Materials Processing*, 251–264. https://doi.org/10.1016/b978-0-08-096532-1.01010-4
15. ASTM F2792-12a. (2015). Standard terminology for additive manufacturing technologies. Retrieved April 26, 2017 from astm.org.
16. 8th international symposium on superalloy 718 and derivatives: Novel processing methods. Retrieved April 26, 2017, from programmaster.org.
17. Tamayo, J. M. G., Riascos, M., Vargas, C., & Baena, L. M. (2021). Additive manufacturing of Ti6Al4V alloy via electron beam melting for the development of implants for the biomedical industry. *Heliyon, 7*(5), e06892. https://doi.org/10.1016/j.heliyon.2021.e06892
18. Carlota, V. (2019, Oktober 7). The complete guide to electron beam melting (EBM) in 3D printing. *3Dnatives*. Geraadpleegd op 1 mei 2022, van www.3dnatives.com/en/electron-beam-melting100420174/#

19. www.thomasnet.com/articles/custom-manufacturing-fabricating/electron-beam-additive-manufacturing-ebam-3d-printing/

20. Mikahila, L. (2021, November 1). Material jetting vs. binder jetting: Which jetting process should you choose? *3Dnatives*. Geraadpleegd op 1 mei 2022, van www.3dnatives.com/en/material-jetting-vs-binder-jetting-300920216/

21. Mirzababaei, S., & Pasebani, S. (2019). A review on binder jet additive manufacturing of 316L stainless steel. *Journal of Manufacturing and Materials Processing, 3*(3), 82. https://doi.org/10.3390/jmmp3030082

22. Introduction to Binder Jetting 3D Printing. (2022). *Hubs*. Geraadpleegd op 5 mei 2022, van www.hubs.com/knowledge-base/introduction-binder-jetting-3d-printing/

23. Introduction to Material Jetting 3D Printing. (2022). *Hubs*. Geraadpleegd op 5 mei 2022, van www.hubs.com/knowledge-base/introduction-material-jetting-3d-printing/

24. Cook, B., Tehrani, B., Cooper, J., Kim, S., & Tentzeris, M. (2015). 8-Integrated printing for 2D/3D flexible organic electronic devices. *Handbook of Flexible Organic Electronics*, 199–216. https://doi.org/10.1016/b978-1-78242-035-4.00008-7

25. AMFG. (2020, April 29). A comprehensive guide to material jetting 3D printing. *AMFG*. Geraadpleegd op 1 mei 2022, van https://amfg.ai/2018/06/29/material-jetting-3d-printing-guide/

26. Engineering Product Design. (2022, February 23). *What is binder jetting and how does binder jetting work*. Geraadpleegd op 1 mei 2022, van https://engineeringproductdesign.com/knowledge-base/binder-jetting/

27. Engineering Product Design. (2022, February 23). *What is material jetting and how does material jetting work?* Geraadpleegd op 1 mei 2022, van https://engineeringproductdesign.com/knowledge-base/material-jetting/

28. Sireesha, M., Lee, J., Kranthi Kiran, A. S., Babu, V. J., Kee, B. B. T., & Ramakrishna, S. (2018). A review on additive manufacturing and its way into the oil and gas industry. *RSC Advances, 8*(40), 22460–22468. https://doi.org/10.1039/c8ra03194k

29. Drop-on-Demand 3D Metal Printing. (2021). *Forcyst*. www.forcyst.com/drop-on-demand-dod#:%7E:text=Drop%20On%20Demand%20is%20a,one%20layer%20at%20a%20time.

30. Burgués-Ceballos, I., Stella, M., Lacharmoise, P., & Martínez-Ferrero, E. (2014). Towards industrialization of polymer solar cells: Material processing for upscaling. *Journal of Materials Chemistry A, 2*(42), 17711–17722. https://doi.org/10.1039/c4ta03780d

31. OneMonroe. (2020, September 19). *What is a desktop 3D printer?* https://monroeengineering.com/blog/what-is-a-desktop-3d-printer/#:%7E:text=A%20desktop%203D%20printer%20is,printers%20are%20small%20and%20compact.

32. What Is a Desktop 3D Printer? (2022). *Raise3D: Reliable, industrial grade 3D printer*. www.raise3d.com/academy/what-is-a-desktop-3d-printer/

33. Alex, M. (2021, June 1). The 4 types of FFF/FDM 3D printer explained (Cartesian, Delta, Polar). *3Dnatives*. www.3dnatives.com/en/four-types-fdm-3d-printers140620174/

34. Fossett, K. (2021, October 10). DeltaBot 3D printer. *WhiteClouds*. www.whiteclouds.com/3dpedia/deltabot/

35. Lin, J., Luo, C. H., & Lin, K. H. (2015). Design and implementation of a new DELTA parallel robot in robotics competitions. *International Journal of Advanced Robotic Systems, 12*(10), 153. https://doi.org/10.5772/61744

36. 3D Printing Technology – Delta versus Cartesian. (2021). *Tractus3D*. https://tractus3d.com/knowledge/learn-3d-printing/3d-printing-technology-delta-versus-cartesian/

37. Stevenson, K. (2020, July 23). A curious thing about 3.00 vs 1.75mm 3D printer filament. *Fabbaloo*. www.fabbaloo.com/2015/07/a-curious-thing-about-300-vs-175mm-3d-printer-filament

38. 3D printers components—how 3D printers work. (2020). *Solectroshop.com*. https://solectroshop.com/en/blog/3d-printers-components-how-3d-printers-work-n40

39. Flynt, J. (2018, August 10). Parts of a 3D printer. *3D Insider*. https://3dinsider.com/3d-printer-parts/
40. Hanna, K. T. (2021, September 30). Open source. *WhatIs.Com*. www.techtarget.com/whatis/definition/open-source
41. Miller, R. R. (2003, April 25). Why do programmers write open source software? *Linux.com*. www.linux.com/news/why-do-programmers-write-open-source-software/
42. Gonzalez-Barahona, J. M. (2021). A brief history of free, open source software and its communities. *Computer, 54*(2), 75–79. https://doi.org/10.1109/mc.2020.3041887
43. Gaudeul, A. (2007). Do open source developers respond to competition? The latex case study. *Review of Network Economics, 6*(2). https://doi.org/10.2202/1446-9022.1119
44. Brasseur, V. M. (2018). *Forge your Future with Open Source*. Pragmatic Programmers.
45. Berman, D. (2022, April 21). GNU general public license | GPLv3 explained. *Snyk*. https://snyk.io/learn/what-is-gpl-license-gplv3-explained/
46. The GNU General Public License v3.0—GNU Project—Free Software Foundation. (2007). www.gnu.org/licenses/gpl-3.0.en.html
47. Myayan Blog. (2021, October 12). What are the advantages and disadvantages of GPL license. *Myayan Blog*. www.myayan.com/advantages-and-disadvantages-of-gpl-license
48. What Are the Pros and Cons of the GPL? (2010, September 27). *Software Engineering Stack Exchange*. https://softwareengineering.stackexchange.com/questions/7720/what-are-the-pros-and-cons-of-the-gpl
49. Zhang, C. (2015). 3d printing, open-source technology and their applications in research. https://doi.org/10.37099/mtu.dc.etdr/62
50. Wasti, S., & Adhikari, S. (2020). Use of biomaterials for 3D printing by fused deposition modelling technique: A review. *Frontiers in Chemistry, 8*. https://doi.org/10.3389/fchem.2020.00315
51. Davis, R. (2021, August 18). What is 3D printing? Definition, technology and applications. *The Engineering Projects*. www.theengineeringprojects.com/2021/06/what-is-3d-printing-definition-technology-and-applications.html#

3 Basic Learning of the 3D Printing Process

3.1. BASIC CONCEPTS OF 3D MODELLING

The preponderance of design websites, periodicals, and multimedia still use 2D visuals even if digital technology is now flourishing. Many aspiring artists and designers are still restricted to this format since it is the quickest and easiest way to visually communicate ideas—from sketching to drawing to rendering.

3D modelling, which involves creating three-dimensional structures using computer programmes, is often used in large-scale production. Usually, computer-aided design specialists are in charge of it. Students who are interested in design frequently shy away from 3D modelling because of its complexity and drawn-out process. Furthermore, technical proficiency is more significant than novelty [1].

3D modelling is used in a variety of applications to produce digital representations of real-world objects. 3D modelling is a subset of Computer Aided Design (CAD), which employs a computer to help any form of a design job's design process. It may be used for a variety of things, but is most frequently used for developing pieces for a portrayal on a computer. The computer model is used to communicate dimensions, material kinds, etc., to anybody viewing the design and can be used to generate control pathways for Computer Numerical Controlled (CNC) equipment. To mimic a real substance, 3D models link a matrix of points in three-dimensional (3D) space using various geometrical elements like triangles, lines, curved surfaces, etc. Since 3D models are collections of data, they can be created manually, automatically (procedural modelling), or by scanning (points and other information). Texture mapping may be utilised to give their surfaces more personality [2].

Although computers could produce 3D models in real time in the past, many video games used pre-rendered images of such models as their sprites. The designer may then examine the model from a variety of perspectives to see if the final output adheres to their original vision for the object. By viewing the design in this context, the designer or company may decide what improvements or adjustments should be made to the final product [3].

1. **Representation:** Two methods can be used to classify almost all 3D models as follows:
   - Solid—These forms specify the volume of the object they represent (like a rock). Solid models are mainly created using constructive solid geometry and are mostly utilised in health science applications.
   - Shell or boundary—These depictions emphasise the item's surface or perimeter rather than its size or volume (like an infinitesimally thin eggshell). The vast majority of visual models used in movies and video games are shell structures [4].

DOI: 10.1201/9781003349341-3

All this solid and shell modelling may be used to create objects that are functionally comparable. The techniques used to build and update them, as well as use standards in different fields and various kinds of assumptions between the model and reality, are the primary differences between them.

Shell modelling should be varied in order to be applicable as a genuine object (i.e., without holes or cracks in the shell). In a shell model, a cube's bottom and top surfaces must have the same thickness, and the first and final printed layers must be error-free. The bulk of images are properly polygonal meshed (and to a lesser extent, subdivision surfaces). When representing deforming surfaces that undergo many topological changes, such as fluids, level sets are an efficient tool.

The act of changing coordinate representations of items, such as the location of an object's centre and a point on its sphere, into polygonal depictions of those same entities is known as tessellation. In polygon-based rendering, objects are converted from abstract representations (or "primitives") like spheres, cones, and other geometries to meshes, which are collections of linked triangles. Triangular meshes are widely used in place of square meshes due to their popularity and ease of rasterisation, for example (the surface emphasis by each triangle is planar, so the projection is always convex). Since not all rendering algorithms employ polygon representations, the tessellation stage is not always required when going from this type of abstract representation to a displayed scene [5].

3.2. TYPES OF MODELLING

Solid, wireframe, and surface 3D modelling are the three main types utilised in CAD. Both are good and bad. There are additional variations, for sure, but the majority of them are either a mix of these three or highly tailored to their individual purposes.

- **Solid modelling**: This is a technique for creating three-dimensional objects. Despite having different shapes, they work together as a set, much like building bricks. Some of these blocks add material while others take it away, depending on the different inputs. Some programmes have the ability to alter solids as though they were being physically milled in a shop using modifiers. Solid modelling is relatively user friendly and computationally straightforward.
- **Wireframe modelling:** It can be helpful when the surface is complicated and curved. You'll soon find that solid modelling's basic building blocks are too complex for some applications, but wireframe modelling provides the accuracy for more intricate structures. But when complexity rises, some drawbacks are exposed
- **Surface modelling**: The intricacy of this modelling procedure has risen. The clear surfaces and complete integration that exceptionally competent applications seek can be managed by more sophisticated programmes that require more effort and processing capacity. However, with this method, you can create forms that are nearly difficult to create using the other two methods [6].

3.2.1.   3-Dimensional Modelling

By arranging several perspectives of the model on a drawing sheet, identifying these views with measurements, and including annotations, a schematic design may be created in three dimensions. If the 3D model below is a solid or surface model, automated hidden-line elimination may be employed. It is necessary to model geometry as a collection of lines as well as other curves, surfaces, or objects in space in order to simulate three dimensions. As was already mentioned, drawings are produced utilising a two-dimensional coordinate system. Three-dimensional (3D) models are constructed in 3D space, frequently utilising a right-handed Cartesian coordinate system. The model construction process can potentially benefit from the usage of a movable world coordinate system (WCS). The initial specification of the model is often done using a fixed coordinate system called the global coordinate system (OCS).

3.2.1.1.   Right Hand Rule

This is a formula for determining the axes' position in a right-handed Cartesian coordinate system. The thumb represents the x-axis, the pointer finger represents the y-axis, and the middle finger represents the z-axis when the right hand's thumb, index finger, and middle finger are placed at triple right angles. Most contemporary CAD programmes use right-handed Cartesian coordinate systems natively, which appears to be the industry norm. The same logic applies to left-handed Cartesian coordinate systems; they merely utilise the left hand. The most common use for them is in espionage, even though they correspond well with flat projections of the longitudinal and latitudinal matrices of orthogonal projections of the crust of the earth. The right-hand rule also provides a straightforward method for determining the direction of the x-, y-, and z-axes. The backside of your hand should be facing the screen while using the default WCS, with your thumb pointed to the right and your index finger pointed upward. Your middle finger is now stretched, and you might see a positive movement on the z-axis. In the WCS, the z-axis is oriented positively in your direction [7].

3.2.1.2.   Using X, Y, Z Point Coordinates

Up to this point in the chapter, the coordinate systems were 2D coordinates that described the locations of point locations on the X, Y planes of the World Coordinate System. The World Coordinate System (WCS), which is the constant coordinate system used by default in CAD programmes like AutoCAD, is a common coordinate system. The WCS's z-axis is parallel to the plane, but its x- and y-axes are both horizontal. A 2D absolute point coordinate is defined by the values X, Y. In a 2D coordinate statement, 0 is the minimal z-axis number. This makes the position specification 3,3 equivalent to 3,3,0. To specify a 3D point position outside of the X, Y plane, a nonzero z-axis integer must be given (Figure 3.2). 3, 3, 3, are the coordinates of this place. Keep in mind that the point is situated 3 units just after the positive x-axis, 3 units all along the positive y-axis axis, and 3 units solely on the positive z-axis.

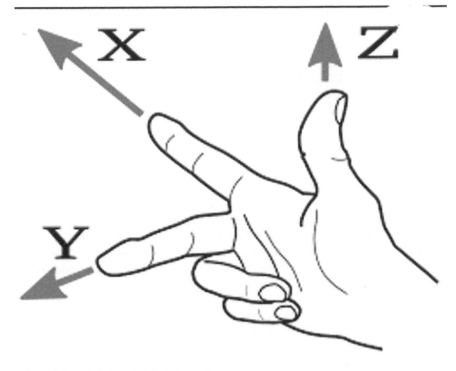

FIGURE 3.1   Right Hand Rule [7].

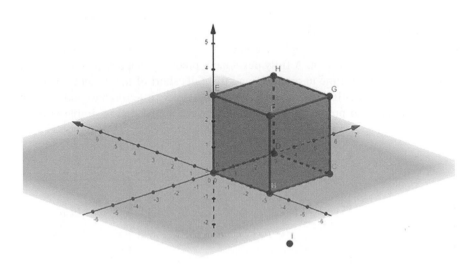

FIGURE 3.2   XYZ Coordinate.

Process of point coordinate modelling:

1. Click Home tab >Modelling panel >Box. >Find
2. Specify the first corner of the base.
3. Specify the opposite corner of the base.
4. Specify the height.
   Command: L or Line
   Specify first point: 3
   Specify second point: 3
   Specify height: 3

3.2.1.3.   Solid Modelling

The solid model, in contrast to wireframes and surface portrayals, which primarily contain geometrical data, incorporates topological data connected to geometric data to accurately and thoroughly show the item. A strong model facilitates accurate design and improves CAD/CAM projects like CIM. The automation of the production process is enhanced through flexible manufacturing. Then, the question of what geometry and topologies are arises.

- Geometry: It relies on the size and measurements of the structural body as well as the precise placement of the object in the chosen coordinate frame.
- Topology: It contains data that may be integrated, such as neighbourhood specifics, connectivity, and associative data. Additionally, this lacks relationship information.

3.2.1.4.   Wireframe Modelling

Wireframe modelling is essential when comparing a 3D graphic model to its source. It allows the designer to view the standard through the model and line up the vertex and edge points with the right reference. Wireframe modelling facilitates rapid and easy concept exposition. A fully developed, precisely mapped prototype for an idea can take a lot of time to produce, and if it falls short of the project's goals, all that time and work is lost. Wireframe modelling makes it feasible to condense the extensive labour and provide a very straightforward model that is simple to create

FIGURE 3.3   Same Geometry But Different Topology.

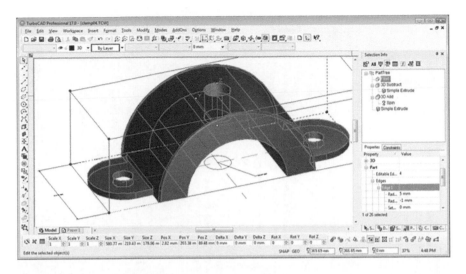

FIGURE 3.4 Solid Modelling.

and simple for others to understand. The overall model is composed of points, lines, curves, circular and conics. Inside a 3D wireframe model, an item is not solidly represented. Instead, the vertices that make up an object's perimeter or the intersections of its edges are represented by a collection of points and how they are related [8].

A few key advantages of 3D wireframe modelling include:

- Simple design development and iterations for entity evaluation.
- A 3D view of the built-in model from any aspect or perspective.
- It enables you to carefully inspect the object's composition, including distances, perspectives, potential variations, and corners, among other things.
- Orthographic and auxiliary perspectives can be automated with the use of 3D wireframe modelling.
- You may inspect the reference geometry for a 3D solid, mesh, and surface model by using this 3D modelling approach.

The disadvantages of 3D wireframe modelling are similar to its benefits. The presentation of inside and outside object surfaces is a problem for designers. Although wireframe modelling makes 3D models generally more understandable, they are still fairly complicated. Furthermore, by providing one-dimensional vision, it enhances the object comprehension [9].

3.2.1.5. Surface Modelling

Surface modelling may be used to produce a visual representation of an asteroid's exterior and contours. Or, to put it another way, it's a platform. These items might be complex biological forms like animals or engine-related mechanical components. Irrespective of the product you are producing, surface modelling requires that you define the external curvature and contours of your stuff.

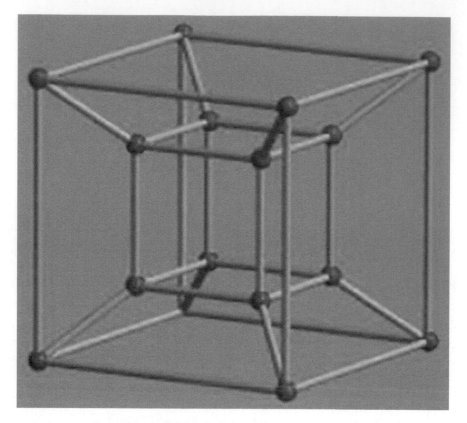

FIGURE 3.5   Wireframe Modelling.

Unlike solid modelling, your item could be practically and geometrically incorrect since it lacks standard mass and thickness specifications. As a result, the designer has the exclusive authority to change the model in a way that solid models are unable to. However, because the object is effectively hollow, this also suggests that surface models cannot be sliced apart like their solid counterparts. Furthermore, bear in mind that all these objects' interfaces may be depicted using polygons or NURBS depending on the specific application [10].

The smooth plane, which can be described by two parallel straight lines, three points, or a line and a point, represents the most basic sort of surface. An illustration of a plane presented by a commercial CAD system. Other instances of types of surfaces frequently used in CAD are shown in Figure 8.

- A tabular cylinder, which is created by producing a curve and projecting it along a vector or line.
- A ruled surface that is created by using edge curves or linear interpolation between two separate normals. Extending that straight line while having its end points lay on the edge curves produces the illusion of a sheet.

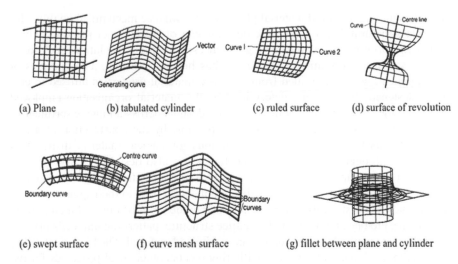

(a) Plane   (b) tabulated cylinder   (c) ruled surface   (d) surface of revolution

(e) swept surface   (f) curve mesh surface   (g) fillet between plane and cylinder

FIGURE 3.6   Various Types of Surface Modelling [8].

- A surface of revolution, that has been formed by rotating a producing curve about just a centerline or vector. When modelling rotated or axially symmetrical pieces, this surface is especially useful.
- A swept surface, which is somewhat of an extension of the surface of revolution, sweeps the identifying curve over a non-circular arc (Figure 3.6).
- A surface with curves or sculpting. A series of generating curves or two families colliding in a crisscross pattern to form this surface type, which is among the most generic of the surface types, results in a matrix of linked surface patches.
- A fillet surface is indeed a barrier that connects two other edges in a smooth transition. It is similar to the fillet arc in curvature creation (generally of a constant, or smoothly changing, radius of curvature).

3.3. DESIGN STRATEGIES

A way for 3D-printed components to enable required functionality that goes beyond just manufacturing a solid part is provided by design techniques that are application agnostic. The design tactics include the following: Architected materials, Stimuli Responsive, Multiple material combinations, Functionally Graded Materials, and Customization. These strategies are described in more detail below. Because they allow for more specialised functionality and property refinement for created devices beyond the choice of materials and printing techniques, these layout concepts are advantageous for medical applications.

3.4. ARCHITECTED MATERIALS

Architected substances are constructed structures with a regular arrangement of subunits, such as a lattice with unit cells obeying a specified topological allocation that makes use of structured material placement to enhance mechanical qualities for a specific structural density. With the development of 3D polymer printing, which enables the printing of complicated geometries with high resolution and precision, the synthesis of architected materials has become efficient. The qualities of the component, which are scaled from the basic material properties utilised to create the designed structure, are determined by the material organisation employed throughout the structure. The ability to create designs across a huge assortment of elastic modulus and density values is made possible by the construction of architected materials by taking into account various topological material distribution techniques. Although bending-dominated foams are also desired because of their energy absorption capabilities, squeeze beam-based lattice structures are frequently employed for transporting loads in medical applications due to their great mechanical efficiency. By first creating a single unit cell made up of beams and then putting unit cells next to one another to build a lattice structure, patterning unit cells offers a straightforward method of constructing an architected material. The biological utility of the architected material, such as facilitating mechanobiological processes for tissue formation, is also influenced by the beam diameter and the topology of the beams inside a unit cell. Hierarchical methods provide a more advanced method of creating architected materials for better mechanics. Due to the bigger holes generated by the hierarchy, hierarchically architected materials have a somewhat lower elastic modulus and a much higher capacity for transporting nutrients, potentially improving the performance of tissue scaffolds used in applications of regenerative medicine [11–16].

3.5.   STIMULI-RESPONSIVE

A synchronized assembly of printed components with guided state changes in response to an external stimulus, such as light, temperature, or pressure, is the foundation of stimuli-responsive designs. The use of external stimuli modifies the system's energy and causes it to drive a desired mechanical motion. Combining contrasting materials with varying responses to a stimulus is a popular technique for stimuli-responsive components. A mixture of shape memory polymers forms a self-folding box, the combined material responses throughout the system provide a guided response. In order to close the box, thermal energy was employed to cause shape changes based on the time-dependent behaviours of each polymer. Moreover, it is possible to ingeniously disperse a single material across a system such that it responds with shape memory and takes on many shapes in response to stimuli. The mechanics and interactions of the materials dictate how the form of the part changes and how long it takes for reactions to occur in stimuli-responsive material design. A glassy polymer and an elastomer were combined and extrusion printed to create a rod-shaped construct as a recent experiment to study the combination of materials with differing levels of sensitivity to stimuli. In this situation, the glassy polymer was more likely to alter form in response to outside heating stimuli. The outcome showed that thermomechanical behaviour might produce more than 300% of the failure strain by carefully adjusting the stimuli. The production of 3D-architected photo-shape memory alloys has been made possible by the application of high-resolution photocuring and high-contrast microdisplays. One of the experiments offered a new 4DMesh technique that involved shrinking and bending a 4D print to create an undevelopable surface using a thermoplastic controller. The study also proved the print's use in commercial packaging and moulds and confirmed its aesthetic, mechanical, and geometric features.

3.6. MULTI-MATERIAL

More research is being done on multi-material 3D printing, which combines materials with different qualities to improve overall functionality and performance in printed objects. Direct ink writing, material jetting, and fused deposition modelling are a few printing techniques that frequently combine multi-material printing. Either a single nozzle extrusion that prints materials one at a time or a different nozzle for each material are used in multi-material printing. In contrast to single-material constructions, multi-material periodic composites have special mechanical characteristics. An inflexible periodic structure implanted in a hyperplastic material, for instance, has been combined using fused deposition modelling to provide high compliance and a rapid rate of strain recovery. Performance was obtained by distributing the applied load evenly across the periodic structure, which improved the entire mechanical response, thanks to the integrated extremely flexible matrix. Multi-material printing has also been utilised to create anatomically realistic medical phantoms that mimic the mechanical characteristics of genuine tissues. Direct ink writing has already been used to illustrate the possibilities of multi-material 3D printing for a functional and shape-morphing structure. Because of the use of an adjustable negative Poisson ratio for a homogeneous cell structure in multi-material printing, the field of metamaterial printing is also progressing. By printing the beams with flexible and stiff polymers, distinct elastic behaviours of the printed material were exhibited, as opposed to employing geometric characteristics to influence the Poisson ratio. Direct ink writing was used to study fibre orientation in a polymer-fibre composite, which is another area where the multi-material approach has been used. This study suggested using an epoxy-resin-based ink to regulate fibre orientation and showed that it might increase mechanical strength by up to ten times. An effective and quick method of printing numerous materials at once is multi-material printing with multiple nozzles. With the ability to regulate the deposition of each material at the size of individual voxels, a direct ink writing multi-material and multi-nozzle print head can print components for a variety of applications using up to eight distinct types of materials.

3.7. FUNCTIONALLY GRADED

Architectural substances that have been developed with a progressive geometric or material transition across the building are known as functionally graded materials Functionally graded materials prevent the dramatic transformation of mechanical characteristics at interfaces and allow a gradual transition of characteristics. Functionally graded materials, particularly as load-bearing supports, hence reduce stress concentrations over surfaces and offer durability. Functional gradients are also helpful in medical uses because they offer a highly complex diversity of biologically inspired gradients and allow for greater control over fluid flow, mass transport, biodegradation, and mechanical performance, like stiffness, strength, and hardness, throughout a designed structure, which is advantageous for biomedical implants. A functionally graded lattice construction is depicted where the thickness of the beams varies depending on where they are. The structure's distortion nature makes it lightweight and good at absorbing energy. The structure exhibits deformation starting at the layer with the lowest density, followed by a layer-by-layer collapse in sequence, with the exception of the last layers, which collapse simultaneously or very quickly following one another. The density gradient facilitates this layer-by-layer deformation, which offers favourable mechanical responses for applications where unexpected mechanical failures are a problem. For patient-specific fabrications for customised medicine, where the preset pattern matches a specific patient's anatomy, customization allows printing components with geometries updated on a per-print basis. In particular, implant devices are printed in bone tissue engineering depending on the patient's imaging bone shape, which offers a better interface to increase host-bone interaction. Dental implants require such modification in order to achieve optimal fit. The ability to arrange medical procedures using the actual component makes the printed model useful for intricate mandibular reconstruction. Planning with the aid of the 3D-printed model may speed up the process while also enhancing the accuracy and quality of the procedure. Manual customization is frequently laborious owing to the amount of layout choices to consider when fitting a component for a patient, which is why image-based approaches are widely utilised for automated design modification. For example, a patient's CT scan's series of 2D pictures is transformed into a 3D image. Then, using optical image information or a number of other imaging technologies, 3D imaging data is transformed into a virtual 3D surface shape and matched with it. Engineers can also utilise imaging data in conjunction with optimization methods to design 3D printed components that are optimised for a patient's unique geometry and function better than parts made using more conventional manufacturing methods [11–17].

3.8.   HOW TO SLICE A MODEL

The process of converting a 3D model into a series of instructions for 3D printers is known as "slicing." The 3D model is effectively "sliced" into thin sheets or surfaces, and the tool path for each surface to be printed is chosen such that it has the best strength, the quickest printing time, etc.

A slicer application converts a 3D CAD model, which is frequently an STL-format file, into G-code that the printers may subsequently utilise [18]. The three main categories of parameters that can be adjusted in a slicer programme are as follows:

- Print settings: layer heights, shells, infill percentage, and speed.
- Filament settings: its included filament diameter, extrusion multiplier, the temperature of the extruder, and print bed.
- Printer settings: nozzle diameter, print bed shape (L × W), and Z offset.

3.8.1.   SLICING SOFTWARE COMPONENT

Understanding the data flow throughout the 3D printing process is crucial.

It may be divided into two components, similar to other types of software: the front end and the back end, or the GUI and the logic. The front end consists of the elements with which users may immediately interact. Along with the G-code and tool path, the user may also see the CAD in this area. The rear components include algorithms or directives for the code to be executed.

Front End	Back End
GUI	STL Reading
STL Visualization	Slicing Algorithm
G-Code Visualization	G-Code Preparation

Front End: GUI stands for graphical user interface. Every tool required for user interaction with the 3D model is included. Figure 3.8 shows the graphical user interface of Creality Slicer, a free tool for slicing. The 3D model's characteristics may be moved, scaled, rotated, and modified by the user.

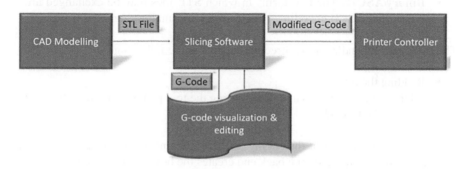

FIGURE 3.7   Data Flow of 3D Printing.

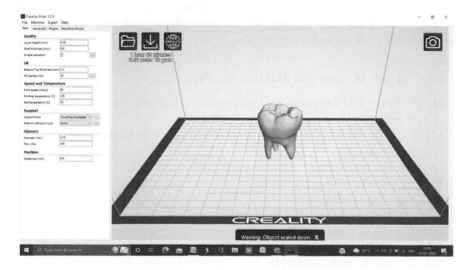

FIGURE 3.8   Creality Slicer User Interface.

The main function of the slicing programme is to work with STL-format files. It preserves the visual representation of the 3D model. It is often referred to as Standard Tessellation Language or Standard Triangle Language. The acronym is "STereoLithography." It includes information about the surface geometry of the model. Simply stated, it encodes the model description so that the slicer recognises it and can generate the graphics. Tessellation, which is associated with STL files, is the process of turning complex 3D structures into 2D models. The following example shows how a collection of small squares or triangles might represent a sphere [19]. Specific tessellation principles help to produce 3D prints of acceptable quality.

- **Angle control:** The spaces in between triangles, which minimize angle change, improve the printing quality.
- **Chordal deviation:** The possibility of a higher distance between both the surface of the structure and the STL extension, known as the chord height, must be preserved. The chord tolerance is typically set between 0.01 and 0.001 mm.
- **Binary/ASCII:** The two forms in which STL files may be exchanged are binary and ASCII encodings. File formats are smaller than ASCII files.

The back end: It handles these three parameters in the following:

- Reading the STL
- Slicing STL into layers and determining the tool path to print the object
- Writing the G-Code

The most important step in the slicing process must include the G-code. Based on the inputs from the receiver, the back end creates the best possible code to create the item. Simply said, a G-code is a collection of commands required to execute a print operation. It assists in sending orders to the machine controller. A 3D slicer produced this G-code automatically. By knowing more about the G-code, one can tackle 3D printing issues. The letters "G" and "M" designate two distinct sorts of instructions. The basic syntax of the code is as follows.

3.8.2. G AND M CODE FOR STARTING THE PRINT

M190 S{print_bed_temperature};Uncomment to add your own bed temperature line
M109 S{print_temperature};Uncomment to add your own temperature line
G21;metric values
G90;absolute positioning
M82;set extruder to absolute mode
M107;start with the fan off
G28 X0 Y0;move X/Y to min endstops
G28 Z0;move Z to min endstops
G1 Z15.0 F{travel_speed};move the platform down 15mm
G92 E0;zero the extruded length
G1 F200 E3;extrude 3mm of feed stock

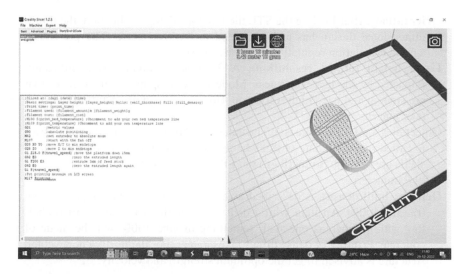

FIGURE 3.9 G-Code Interface.

G92 E0;zero the extruded length again
G1 F{travel_speed}
;Put printing message on LCD screen
M117 Printing . . .

G and M Code for ending the print:

M104 S0;extruder heater off
M140 S0;heated bed heater off (if you have it)
G91;relative positioning
G1 E-1 F300;retract the filament a bit before lifting the nozzle, to release some
   of the pressure
G1 Z+0.5 E-5 X-20 Y-20 F{travel_speed};move Z up a bit and retract filament
   even more
G28 X0 Y0;move X/Y to min endstops, so the head is out of the way
M84;steppers off
G90;absolute positioning
M81
;{profile_string}
Where, G and M = Command Type
000 = Command Number
<par1><par2><par3> = Command parameter

3.8.3. FEATURES OF A SLICING SOFTWARE

After obtaining the STL extension from the CAD application, there are a few software settings that you must handle.

Orientation: After loading the STL file into the slicing application, the three-dimensional design has to be oriented to the free space. The previously agreed orientation regulation must be followed.

Layer height (object resolution): A 3D object's quality is significantly impacted by the layer thickness. Modest layer heights create surfaces that are smooth and have high resolution. However, the entire printing time can take longer.

Object Shells (Shell Thickness): The outermost layer that forms the item's wall is called the shell. The object's hardness varies depending on how many shells are present. Strength grows as the number of shells does!

Infill Parameter:

Density: Infill density is described as a percentage. It could range between 0% and 100%. An item with a density of anything like 0% will be hollow within, whereas one with something around 100% will be made of concrete. Printing of dense things takes longer. For the infill, a honeycomb structure is advised to get the optimum strength and velocity.

3.9. SUPPORT STRUCTURES

In procedures for additive manufacturing, one layer is produced on top of an already-existing layer. According to this, each layer requires an underlayer to sustain it. If part of a design hangs over the edge, it still needs a supporting framework. This supporting layer is referred to as the support structure. When it is required, supports can be added utilising slicer software. The support structures are easily reusable once they've been printed. Figure 3.10 shows a component with a blue hue, which represents the support structure for the 3D model that is being presented.

Rafts and brims: 3D-printed objects frequently become trapped on respective print beds (physically, not in software). The slicer provides some helpful strategies to prevent this. It takes advantage of a few structures that facilitate separation.

- **Brim** (as shown in Figure 3.10): The layer that contacts the real 3D item is one that was printed all along printing platform.
- **Raft** (as shown in Figure 3.11): Before printing the design, a number of layers are produced. It acts as the 3D asteroid's removable basis.

Print speed: Your 3D model's form will have an impact on the slicer software's speed settings. The slicer's speed is set by default, which may be slower than you want. Depending on the drugs you use, the pace may also fluctuate. For PLA or ABS, the recommended print speed is typically 40–60 mm/s. The following is a summary of the print speed considerations that must be made based on the form of your model.

Print rates are really split into two categories: print motions and non-print movements.

Optimized Print Paths: The printing route has to be optimised for 3D printing to work properly. It affects factors such as printing efficiency, surface grit, and strength. The printing pathways have a direct impact on how long

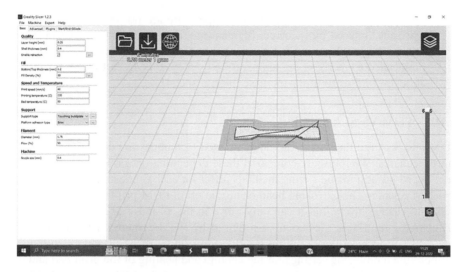

FIGURE 3.10 Brim Support Structure.

FIGURE 3.11 Raft Support Structure.

it takes to complete a given print. A route planning approach may be used to optimise the printing process as a whole. Fabrication will go faster as a result. In the slicer application, there are pre-defined print routes from which to pick. Some of the most well-liked paths are provided by slicer software like Cura and Slic3r.

Temperature: The most important slicer parameter is variable temperature. In line with the print speed parameters, it should be corrected. It is essential to increase the temperature if the printing speed is adjusted to a higher value in any way. If the print speed and temperature are not balanced, "under extrusion" caused by incorrect filament melting may result. Furthermore, hot weather might result in "excessive extrusion" and blobs or zits. It is recommended to increase the temperature by 5 to 10 degrees Celsius for just about every 5 to 10 mm/s increase in print speed.

3.10. FDM VS SLA SLICING

FDM and SLA are the two most widely utilised techniques in additive manufacturing procedures. We routinely compared these depending on a range of characteristics. Let's look at how the slicing methods differ for each of these [20].

FDM Slicing: The extrusion head or nozzle of an FDM printer can move in the z-axis to change layers, but it can only move in the x- and y-axes during printing. To operate, multi-axis movement is required by FDM printers. This needs to be taken care of to have the optimum printing uniformity. FDM printers also require temperature monitoring settings for the cooling and heating of the print bed and the extruder. Two of the most effective slicer programmes for FDM printers that have been evaluated to execute pre-defined methods to execute the G-code are Cura (free) and Simplify3D (paid).

SLA Slicing: During SLA printing, the construct plate moves up and down to dip into the resin chamber and create layers one at a time. Therefore, the movement for SLA printing would be along the z-axis. SLA and FDM slicing are fundamentally different from one another since the output file for SLA printers isn't truly a G-code file. Slicing software is pre-installed on SLA printers. Since SLA printers lack extruders, the nozzle temperatures in this instance are useless. Instead, some of the crucial factors include exposure duration and lifting speeds. Two third-party programmes that are often used with SLA printers are ChiTuBox and FormWare.

3.11. CONCLUSION

This idea of slicing software is essential to 3D printing. How accurate your 3D models are depends heavily on the slicing process. It would be simpler for you to manage the 3D printing process if you are aware of several slicing-related ideas. 3D model slicing is the last stage in the three-dimensional printing process. Your 3D models' role in the process will make evident the 3D printing criteria you set for them. We made an attempt to provide you a fundamental overview of slicing software so you could dispel any lingering questions.

3.12. WHAT IS G-CODE?

G-code is the name of the most widely used CNC machine computer language in computer-aided design and engineering (CAD/CAM) (also known as RS-274).

G-code provides metric-based numerical control to CAM-controlled equipment, such as CNC milling machines. The perfectly alright control enabled by G-code and other CNC languages enables the accuracy for additive and reduction-based manufacturing using a range of materials.

The computer that controls the motors is given instructions via the G-code and other CNC codes used in manufacturing equipment on how far and how rapidly to move. For instance, the spinning tip of a mill may form a metal block into an intricately mechanical part. G-code can provide the instructions required for the computer-controlled equipment to move the head through 3D moves at different speeds to create a camshaft.

In the early 1950s, MIT developed the first numerical computer control language. The Electronic Industries Alliance initially standardised the original version of G-code in the 1960s. The exact, repeatable manufacturing capabilities offered by G-code and similar languages are used nowadays in the creation of various sorts of consumer goods, particularly in scientific instruments.

To have a deeper understanding of 3D printing, you must, however, grasp the principles of G-code. With this information, you will be able to modify the firmware for 3D printers like the RepRap and Marlin and manage print processes much more efficiently [21].

Each G-code command line adheres to a certain syntax. Because there is only one command per line, this might result in extremely long codes.

Any given line's first parameter is always the command code itself. As we've seen, it can be either a "G" or "M" code type, then a command-specific number. For instance, the instruction "G0" denotes a linear motion.

The command is then further defined by its arguments. These parameters for a G0 linear move comprise the destination position and the velocity, both denoted by capital letters.

G0 and G1: The G0 and G1 commands both perform linear movements. By convention, G0 is used for non-extrusion movements like initial and travel moves, while G1 encompasses all the extruding linear motion.

Both commands function the same, though. The parameters for G0 or G1 include the final positions for *all the x-, y-, and z-axes*, the amount of extrusion to be performed during the move, and the speed, specified by the feed rate in the set units.

So let's start with an example:

G1 X90 Y50 Z0.5 F3000 E1 tells the printer to move in a straight line (G1) towards the final coordinates. X = 90 mm, Y = 50 mm, Z = 0.5 mm at a feed rate (F) of 3,000 mm/min while extruding (E) 1 mm of material in the extruder.

Most linear moves are performed within a single layer, meaning that the Z coordinate is usually omitted from the command line.

3.12.1.   G90 and G91: Absolute and Relative Positioning

The G90 and G91 commands tell the machine how to interpret coordinates used for movement. G90 establishes "absolute positioning," which is usually the default, while G91 is for "relative positioning."

Neither command requires any parameters, and setting one automatically cancels the other. The way positioning works is quite simple, so let's jump right in.

Let's say we want to move the printhead to X = 30 in a line. In absolute positioning mode, that would look like this:

G90; sets absolute positioning G0 X30; moves to the X = 30 coordinate

This simple move would tell the printer to move the printhead so that it's positioned at X = 30. Now, for a relative positioning move, we need to know where the printhead is currently. Let's assume it's at X = 10:

G91; sets relative positioning G0 X20; moves +20 mm along the x-axis
G91 first tells the machine to interpret the coordinates as relative to the current
        position (X = 10). Knowing that, the machine simply needs to move 20 mm
        in the x-axis positive direction, thus reaching X = 30, as we'd like.

G28 and G29:

We call "homing" the process of setting the physical limits of all movement axes.
        The G28 command will perform this task by moving the printhead until it
        triggers *end-stops* to acknowledge the limits.

Homing is important not only for the machine to orient itself but also to prevent the printhead from moving outside the boundaries. The G28 command is usually performed before every print process.
        Another command, G29, starts the *automatic bed leveling* sequence. There are many different *methods for leveling a bed* prior to printing, as this is usually set by *firmware* and not by the final users. For this reason, we won't get into details surrounding the methods and command parameters. Just know that G29 is usually sent after an auto-home (G28) and should perform the automatic bed leveling as *determined by the firmware*.

EXAMPLE

        G28 X Y; home the x and y-axes only
        G28; home all axes

Specific axes can be individually homed by including X, Y, or Z as parameters. Otherwise, G28 alone will home all three.

        G29; perform automatic bed leveling sequence

If you want to run an auto bed leveling sequence, remember to send G29 after performing the homing process.

3.12.2. M104, M109, M140, AND M190: SET TEMPERATURE

These are essential miscellaneous commands, which, again, don't involve any motion.
        To start, the M104 command sets a target temperature for the hot end to reach and keep it until otherwise instructed.
        Some of the parameters include the actual temperature value (S) and which printhead (T) to heat (for multiple extrusion setups).

EXAMPLE

M104 S210; set target temperature for hot end to 210 degrees

This command line instructs the machine to heat up its hot end to 210 °C and assumes there is only one hot end in this extrusion setup. After setting the target temperature, the printer will go on to perform the next command line while heating the hot end.

Alternatively, if we wanted to wait until that target is reached before moving on to the next line, we can use the M109 command.

M109 S210; set target temperature for hot end to 210 degrees and do nothing until reached

Setting the bed temperature is very similar to the hot end, but instead with the M140 and M190 commands:

M140 S110; set target temperature for bed to 110 degrees
M190 S110; set target temperature for bed to 110 degrees and do nothing until reached

3.12.2.1.   Phase 1: Initialization

Any program's initial section lists the steps that must be taken in order to get ready to start printing the model. The very first six lines of startup G-code from a real 3D printing process are shown shortly.

M190 S50.000000
M109 S220.000000;
Sliced at: [date] [time];
Basic settings: Layer height: 0.25 Walls: 0.4 Fill: 10;
Print time: 27 minutes;
Filamen used: 0.76m 1.0g

We now understand that the first line instructs motions to utilise absolute positioning and the second line instructs the extruder to interpret extrusion in the same manner.

The bed and nozzle begin to warm up to the desired temperatures thanks to the third and fourth lines. Because it won't be waiting for the desired temperature, the printer will automatically home and balance the bed as it heats up.

Some initialisation procedures (like the one used by PrusaSlicer) include cleansing the nozzles, which involves printing a single straight line before beginning the printing procedure.

3.12.2.2.   Phase 2: Printing

The magic takes place here. You can see that it is hard for us to discern what the nozzle is actually doing if you glance at a sliced G-code data.

Since 3D printing is a layer-by-layer process, you'll notice that this phase involves several motions in the XY plane even if only a single layer is being printed. Once it is complete, the start of the following layer will be defined by a very slight movement in the Z direction.

3.12.2.3. Phase 3: Reset the Printer

When printing is finished, a few last lines of G-code instructions return the printer to a sensible default condition.

The hot end and bed warmers could be switched off, the nozzle could move to a predetermined point, the motors could be deactivated, and so on.

3.13. MATERIAL CONSIDERATION

Over the past ten years, technology advancements and associated material advances have drastically altered how scientists, designers, and manufacturers focus on and rely on 3D printing throughout development and deployment.

Each additive manufacturing method modifies the material using heat, light, or other concentrated energy. Photopolymers, powdered thermoplastics, filament thermoplastics, and metals are four fundamental material types that relate to many technologies. Because technology and materials are usually linked, the optimum strategy for a given project will often depend on the resources it needs.

The best material choice is influenced by a variety of factors. To select a material that satisfies the profile and recalls every aspect a part's material must have is a difficult task. So, this essay aids everyone in making the greatest decision. By using this comprehensive tool to search through a range of characteristics and properties, you may select the material that will best meet the demands of your project [22].

3.13.1. PLASTIC

Polymer is the most widely used raw material for 3D printing at the moment. One of the most varied materials for 3D-printed toys and home furnishings is plastic. This process is used to create items like action figures, vases, and desk accessories. Polymer filaments are offered on spools and come in both matte and glossy finishes. They are readily available in clear form as well as vibrant hues, with red and lime green being particularly well-liked. The appeal of plastic is simple to grasp given its hardness, flexibility, smoothness, and vibrant array of colour possibilities. Plastic is often inexpensive, making it a good choice for both producers and consumers.

In FDM printers, thermoplastic filaments are typically melted and, layer by layer, moulded into shape to create plastic items. Typically, one of the following ingredients is utilised to create the many forms of plastic used in this process:

- **Polyastic acid (PLA):** Polyastic acid, one of the greenest additives for 3D printers, is derived from organic materials like sugar cane and corn starch and is thus biodegradable. In the upcoming years, polyastic acid-based polymers, which are readily available in both soft and hard forms, are anticipated to rule the 3D printing market. Hard PLA is more durable and so better suited for a wider range of items.
- **Acrylonitrile butadiene styrene (ABS):** ABS is a preferred material for 3D printers used at home because of its robustness and safety. The substance, which is also known as "LEGO plastic," is made of pasta-like threads that give ABS its stiffness and flexibility. Because ABS comes in a variety of

colours, it may be used for things like stickers and toys. ABS is also used to create jewellery and vases, which is becoming more and more common among artisans.

- **Polyvinyl alcohol plastic (PVA):** PVA is a plastic that may be used for support materials of something like the soluble kind and is commonly used in low asset printers. PVA can be a cheap alternative for objects with a short lifespan even if it is not appropriate for products that need to be highly durable.
- **Polycarbonate (PC):** Polycarbonate is utilised less commonly than the previously mentioned plastic kinds, and it only functions in 3D printers with nozzle designs and high operating temperatures. Polycarbonate is used to create moulding trays and inexpensive plastic fasteners, among many other things.
- The forms and consistencies of plastic objects produced by 3D printers range from flat and spherical to grooved and mesh. An inventive selection of 3D-printed plastic objects, including mesh bracelets, cog wheels, and Incredible Hulk action figurines, may be found by conducting a fast image search on Google. Polycarbonate spools are now available in vibrant colours at most supply stores for DIY crafters.
- **Powder:** Powdered materials are used to build objects on today's more advanced 3D printers. The powder is melted and dispersed in layers on the inside of the printer until the required thickness, texture, and designs are produced. The powders can originate from a variety of items and sources, but the following are the most typical:
- **Polyamide (nylon):** Polyamide enables for great degrees of detail on a 3D-printed object because of its strength and flexibility. The substance is particularly well suited for fusing and interconnecting components in a 3D-printed object. Everything from handles and fasteners to toy vehicles and characters is printed in polyamide.
- **Alumide:** Alumide powder, which is a mixture of polyamide and grey aluminium, produces some of the strongest 3D-printed sculptures. The powder, which may be recognised by its granular and sandy appearance, is trustworthy for industrial prototypes.
- Steel, copper, and other metals are easier to transport and mould into the necessary forms when they are in powder form. Metal powder must be heated to the point where it can be disseminated layer by layer to create a finished shape, just like the many forms of plastic used during 3D printing.
- **Resin:** Resin comprises one of the less popular and more constrained 3D printing materials. Resin provides less flexibility and strength than other 3D-applicable materials. Resin is a liquid polymer that solidifies when exposed to UV radiation. Although resin is often available in transparent, black, and white variations, certain printed things have also been made in orange, green, red, and blue.

The material comes in the following three different categories:

- **High-detail resins:** Used frequently for little miniatures that demand fine detail. For instance, this grade of resin is frequently used to print intricate clothing and face features on four-inch figures.

- **Paintable resin:** The resins throughout this class are renowned for their aesthetic appeal and are occasionally utilised in 3D printers with smooth surfaces. Fairies and other figurines with realistic facial features are frequently constructed of paintable resin.
- **Transparent resin:** Since this kind of resin is the toughest, it is most suited for a variety of 3D-printed objects, frequently used for models that need to be translucent in aspect and soft to the touch.

3.13.2. METAL

Metal is the second-most common material in the 3D printing sector and is produced using a method called direct metal laser sintering (DMLS). Manufacturers of aviation equipment have already adopted this method, using metal 3D printing to expedite and streamline the assembly of component parts. DMLS printers have also grabbed on with producers of jewellery items, which can be made faster and in bigger quantities—all without the many hours of excruciatingly delicate labour—with 3D printing.

Stronger and maybe more varied daily goods can be made from metal. On 3D printers, jewellers have created engraved bracelets out of steel and copper. The fact that the printer does the engraving task is one of the method's key benefits. As a result, bracelets may be created in only a few automatically programmed phases without the need for the manual effort that traditional engraving work needed.

Additionally, the development of metal-based 3D printing is allowing machine makers to employ DMLS to eventually create at rates and in quantities that are not feasible with existing assembly technologies. Supporters of these innovations claim that 3D printing would enable machine manufacturers to generate metal components with strength that is superior to those made of traditional, refined metals.

In the meanwhile, the aerospace sector is advancing the utilisation of 3D components. In what has been the most aggressive effort of its type, GE Aviation aims to manufacture 35,000 engine injectors by 2020.

Similar to the numerous 3D printer plastic types, the variety of metals that may be used with the DMLS technology is broad:

- Stainless steel: This material is perfect for printing out cookware, cutlery, and other goods that can eventually come into contact with water.
- Bronze: Can be used to create fixtures like vases.
- Gold: Perfect for printed necklaces, bracelets, earrings, and rings.
- Nickel: Good material for coin printing. Aluminum is the best metal for thin items.
- Titanium: The material of preference for sturdy, reliable fittings.

Metal is used throughout the printing process in the form of dust. To make the metal dust harder, it is burned. This enables printers to forego casting and create metal components directly from metal dust. These components can then be electropolished and sold when the printing process is finished.

Metal dust has been utilized to create commercially viable objects like jewellery, although it is most frequently employed to print prototypes of metal instruments. Medical equipment has even been created using powderized metal.

When metal dust is used for 3D printing, the process allows for a reduced number of parts in the finished product. For example, 3D printers have produced rocket injectors that consist of just two parts, whereas a similar device welded in the traditional manner will typically consist of more than 100 individual pieces.

3.13.3. CARBON FIBRE

In 3D printing technology, plastic materials are coated with composites like carbon fibre. The goal is to strengthen the plastic. In the 3D printing sector, carbon fibre combined with plastic has been employed as a quick and practical replacement for metal. The significantly slower process of carbon-fibre layup is anticipated to be replaced in the future by 3D carbon fibre printing.

Industries can minimise the procedures needed to construct electromechanical devices by using conductive carbomorph.

3.13.4. GRAPHITE AND GRAPHENE

Due to its strength and conductivity, graphene has grown to be a preferred material for 3D printing. The substance is perfect for flexible gadget components like touch-screens. Photovoltaic arrays and construction components both utilise graphene. Graphene represents one of the 3D-applicable materials that proponents of the option believe is among the most flexible.

The 3D Group and Australian mining firm Kibaran Resources' collaboration gave the usage of graphene in printing its biggest boost. In lab testing, pure carbon, which was initially identified in 2004, has emerged as the substance with the highest electrical conductivity. Since graphene is both light and strong, it may be used for a variety of items.

3.13.5. NITINOL

Nitinol, a substance used often in medical implants, is prized in the 3D printing community for its extreme flexibility. Nitinol, a material made from a combination of nickel and titanium, can bend to great angles without fracturing. The fabric may be made to retake its original form even after being folded in half. Because of its strength and flexibility, nitinol is among the strongest materials. Nitinol makes it feasible for printers to perform tasks that would otherwise be unfeasible while producing medical items [23].

3.14. CASE STUDIES

The worldwide additive manufacturing market has advanced swiftly, providing more extensive and valuable applications. The benefits it provides over traditional manufacturing are what have caused this transformation to accelerate.

The uses for additive manufacturing now seem to be practically endless. Elevated consumer (home, fashion, and entertainment) and industrial (aerospace, medical/dental, automotive, electrical) items are made of it, and improvements in polymeric materials continue to open up new opportunities for the manufacturing sector.

Discover what additive manufacturing is, how it works, and what kinds of polymers are often employed in each step of the production process. Consequently, the next section contains some examples of 3D printing cases:

3.14.1. NASAL PROSTHESIS FABRICATION USING RAPID PROTOTYPING AND 3D PRINTING (A CASE STUDY)

A 22-year-old female patient with such a nasal deformity was sent to the prosthodontics division of the University of Baghdad's college of dentistry. The postoperative nasal deformity was caused by the entire rhinectomy for squamous cell cancer.

The flaw stretched laterally short of something like the nasolabial fold, inferiorly to the nose's base and anteriorly to its root (Figure 3.12). Although she had already

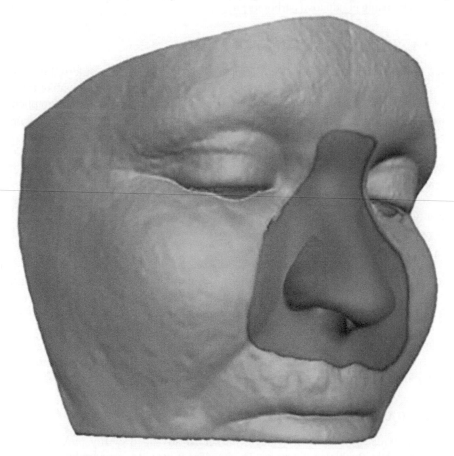

FIGURE 3.12   Scaning Model of Human Nose [23].

undergone nasal plastic surgery to improve her appearance, her primary motivation for seeking a prosthesis was poor facial appearance. The patient didn't have a pre-operative picture that may have been used to create a prosthetic model. Therefore, it was chosen to develop the nasal prosthesis using CAD-RP technology.

A CT scanner was used to perform Computed Tomography (CT) scans of the patient's head (Siemens AG Medical Solutions, Germany); contiguous spiral computer tomography (CT) scans of the person's face were performed at 0.6 mm thickness spacing. A scan was performed from the supraorbital ridge to the intersection of the upper and lower lip after the patient's head had been stabilised. Images created from the obtained data using Digital Imaging and Communication in Medicine (DICOM) were retrieved. Mimics Software was used to transform the collected CT data sets into STL files (Materialise, Leuven, Belgium). Before beginning the actual segmentation, the photos were smoothed gently (decimation factor 0.50, smoothing iterations 20).

The models were not subjected to any flattening or filtering after segmentation. The face's centerline served as the axis of symmetry. The defect's length and breadth were measured.

3.14.1.1. Parametric Model Development

There are several anatomical components to the human nose that seem to be the same in all people. Each nose is distinctive, nevertheless, due to variations in angles, depth, and protrusion. A nose parameterized model was created with completely changeable anatomical structures depending on these common traits.

An easy-to-create model of a prosthetic nose was mostly made using the Zbrush sculpting programme (Pixologic Inc., USA).

To make a comparable prosthesis, this simple nose model was placed over an STL file created from a patient's head soft tissue CT scan.

After conversion, the completed model was entered into 3-matic Medical 11.0 (Materialise, Leuven, Belgium) for examination and airtightness. The design is then sent to the AutoMaker programme for G-code generation and cutting in order to create RP. The 3D printer (RBX01CEL Robox 3D Printer, UK) prints the nose's final product in ABS filament, which is subsequently polished with sandpaper. Any pressure areas were released while the process was carried out [24].

The ABS material was burnt off in the oven after the printed sculpture was invested with investment material together into mould. To mimic the characteristics of skin, maxillofacial silicone elastomer (Factor II, Inc., USA) was utilised. Incorporating intrinsic pigments (Factor II, Inc., USA) into the silicone to match the patient's skin tone was done in accordance with the manufacturer guidelines. The mould was then filled with silicone.

The silicone prostheses is taken from the mould after 24 hours, and then it is polished and finished to have a smooth texture.

The silicone prostheses is painted with extrinsic pigments (Factor II, Inc., USA) to mimic the texture of the skin's surface. After that, the nasal prosthetic was placed into the flaw and secured with medical glue.

It may be quite devastating to lose any part of the body, especially one from the face. Since it is situated in the most noticeable centre region of the face, the nose

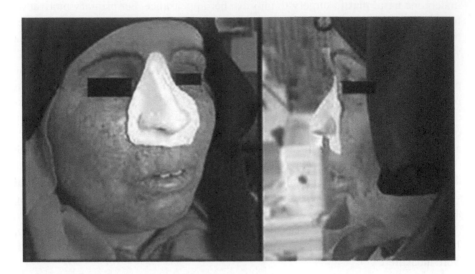

FIGURE 3.13 Nose Implantation [23].

deserves considerable attention. Additionally, it is a feature that can't be covered up in a casual manner like eyes (with tinted glasses) and ears (behind the hair). Therefore, a prosthesis that precisely replicates the patient's natural nose is essential for the rehabilitation process to be successful.

3.14.2. TRUMPF Uses 3D Printing to Improve Satellites and Aircraft

TRUMPF is now exhibiting at the Paris Air Show, the largest and leading aerospace industry trade show in the world, where it is showcasing how additive manufacturing can enhance satellites and aeroplanes.

Somewhere at Paris Air Show, the largest aerospace industry trade show in the world, TRUMPF is showcasing how additive manufacturing can enhance aeroplanes and satellites. A wide range of ever-tougher standards are applied to satellites. On the one hand, they must be as light as possible since the client must pay several hundred thousand euros for each kilogramme that a launch vehicle sends into orbit. Satellites should, however, be strong enough to survive the extreme forces encountered during launch. "One of the largest industrial consumers of additive manufacturing in the world, the aerospace sector has a market share of over 20 percent. We are steadily expanding our market share and helping to establish the process as a key technology," says Thomas Fehn, TRUMPF's general manager for sales in charge of additive manufacturing (AM). Weight reduction is crucial for aeroplanes since it reduces fuel consumption significantly. This lessens their financial and environmental effect. Due to their ability to produce parts that are both extremely lightweight and durable, additive technologies are the ideal fit for the aerospace sector. These techniques don't add material where it isn't truly needed; in contrast, traditional techniques like milling and casting sometimes struggle to remove extraneous material. Light metals like aluminium and titanium are easily handled by 3D printers, and AM engineers and designers have far more design freedom since they are not constrained by the constraints of conventional manufacturing techniques.

Laser metal fusion (LMF) and laser metal deposition are the two primary techniques used by the aerospace sector, and TRUMPF has experience in both (LMD). LMF is completed totally inside the 3D printer, with a laser, layer by layer, creating the item from a powder bed. For making complicated parts for engines, combustion chambers, specialised aircraft components, and related applications, LMF technology is especially well suited. LMD, or laser metal deposition, employs a laser beam to add layers to a part's surface while injecting metal powder through a nozzle. LMD may also be utilised to create extremely massive components quickly. The creation of prototypes and maintenance of substantial components like gas turbines and compressor blades are typical examples.

Three examples of how 3D printing is improving the aerospace industry:

3.14.2.1. Weight of Satellite Mounting Structure Reduced by 55%

TRUMPF is use the Paris Air Show as a chance to present actual aerospace advanced manufacturing solutions. A 3D-printed anchoring framework for Germany's Heinrich Hertz communications satellite, which will be used to test the space-worthiness of new technologies for communication has been ordered by the space corporation Tesat-Spaceroom GmbH& Co.KG. Strap-on motors are utilised in the mounting framework to modify microwave filters. Engineers were able to optimise the topology of the mounting structure and reduce its weight by 55% while working with the business AMendate. Instead of weighing 164 grammes, the mount now only weights 75 grammes. "This is just one example of how we can use additive processes in satellite construction to reduce weight and increase payload capacity," says Matthias Müller, industry manager additive manufacturing for aerospace and energy at TRUMPF. On TRUMPF's TruPrint 3000 3D printer, the team of specialists created the newly developed component. Conventional techniques cannot be used to create the new geometry. The redesigned mounting construction is stronger as well as lightweight. The new mounting framework can endure the same high stresses and will maintain its shape effective during satellite launch. For the Federal Ministry of Economics and Energy and in collaboration with the Federal Ministry of Defense, DLR Space Administration is carrying out the Heinrich Hertz satellite mission.

3.14.2.2. Cost of Engine Parts Reduced by Three-Quarters

TRUMPF is also demonstrating an AM use case for the aviation industry. The TRUMPF specialists used 3D printing in conjunction with Spanish supplier Ramem to improve a component known as a "rake." This component is used by manufacturers to gauge the engine's temperature and pressure throughout engine development. Performance tests for aeroplanes sometimes include these types of metrics. Rakes are subjected to extremely high pressures and temperatures since they are mounted directly in the engine's air flow. They must adhere to certain dimensions specifications in order to give accurate measurements. Rakes are costly and labor-intensive to make the traditional way. While inserting six delicate tubes, soldering them into place, and enclosing the body of the rake with something like a cover plate, workers create the foundation construction on a milling machine. The rake must be discarded if even one of these tubes is misaligned. TRUMPF used the TruPrint 1000 3D printer to create an optimised rake shape. By redesigning the item in this way, the manufacturer can create it more quickly and use around 80% less material, which lowers the whole cost by 74%. "This result shows that 3D printing can save a significant amount of time, material and money in the aircraft industry," says project manager Julia Moll from TRUMPF Additive Manufacturing [25].

3.14.2.3. Making Engine Blades Easier to Repair

They are also showcasing a few examples of LMD technological applications. These include the LMD repair of an aviation engine's high-pressure compressor blade, also referred to as a 3D aeroblade. During flight, these components must resist significant temperature swings. They frequently exhibit deterioration on the edges and tips due to continual exposure to water and dust. To keep the engines operating efficiently, aviation experts must frequently fix the blades. The LMD approach is ideal for this task. The material in some areas of the blades is only 0.2 millimetres thick. In these sorts of applications, conventional techniques soon approach their limitations. But using LMD technology, the laser's location can be accurately controlled to within a tenth of a millimetre before it delivers a precisely calculated dosage of energy. The mechanism feeds in materials precisely matching the part's composition while simultaneously period. This procedure normally takes a few minutes, based on the application. It makes it simple to fix the blades more than once, thereby lowering the cost per component for each engine maintenance. "Laser metal deposition delivers a low dose of energy – and that makes it perfect for aerospace applications. We can use it not only to repair and coat parts, but also to build up three-dimensional structures. That's simply not possible with conventional welding methods,"

says Oliver Müllerschön, head of industry management laser production technologies at TRUMPF.

3.14.3. CASE STUDY ON MEDICAL INSOLE

According to Zuniga et al., two polymers—thermoplastic polyether-polyurethane and thermoplastic polyurethane polyester-based polymer—were used to create 3D-printed insoles, and a diagnostic and therapeutic protocol and inexpensive electronic system were used to assess how much pressure was placed on the plantar surface during walk tests. The two insoles that were 3D printed responded as well to conventional insoles [26].

3.14.3.1. Materials and Methods

Scanners, models, and 3D printers were chosen as the digital technology advancements for the production workflow in this study. A thermoplastic polyether-polyurethane elastomer with additives, and a thermoplastic polyurethane polyester-based polymer, both polymer blends, were employed to fabricate the insoles. Typically, these materials are taken into consideration while creating a medical device prototype. For this technology demonstrator, a healthy adult male patient was chosen to test the insoles during walk trials and participate in the manufacturing workflow as a client. An FSR-based system was used to objectively assess the distribution of plantar pressure. A procedure derived from Owings et al. [11] and Chambers and Sutherland [12] was used to assess the effects of the insoles. This study was accepted by the Universidad Peruana Cayetano Heredia's institutional bioethics committee (SIDISI: 102533). Only the researchers had access to the database containing the patients' anonymised data, which was secured by a password.

3.14.3.1.1. Foot Scan

In the initial step, a precision hand scanner, a 3D Systems Sense 2 model, was used to create the foot prototype. By putting the patient's foot in a resting posture and rotating around, the scan was carried out. Per foot, the scanning procedure took four minutes. Between 40 and 60 centimetres separated the foot from the scanner.

3.14.3.1.2. Insole Design

The second step of insole modelling was carried out using Gensole software [13]. This programme is a web-based application that enables creating insoles for 3D printing to create them.

The insole's top surface was moulded to resemble the foot model. To improve fit well within shoes, a suitable padding density was selected, and contour curves were applied to the insoles. The software was run with the following settings to create this insole:

- Total insole length: 270 mm.
- Measurements at the ends of the insole: 25 and 10 mm.
- Infill up to 50% in the most extensive contact area, as shown in dark green in Figure 3.14.
- The minimum thickness of the insole was 3 mm.
- Previous steps were performed for the right and left foot.

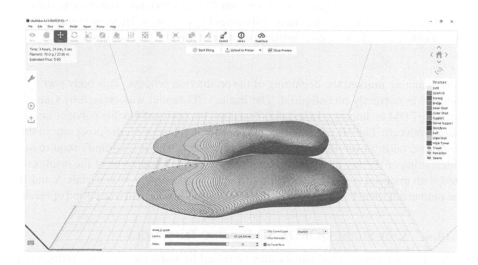

FIGURE 3.14 Medical Insole Model in Slicer Software [25].

TABLE 3.1
Material Comparison.
Color Green Red
Brand FilaFlex NinjaFlex

Material Type	Density (g/cm³)	Hardness Shore A	Elongation at Break (%)	Tensile Strength (MPa)
Thermoplastic polyether-polyurethane	1.08	70	900	30
Thermoplastic polyurethane polyester-based polymer	1.19	85	660	26

3.14.3.1.3. Digital Manufacturing

FDM technology and flexible filament were used for insole fabrication. The properties of these polymers are described in Table 3.1.

Gensole tool-generated AMF format files are required to be translated to G-code in order to be used in 3D printing insoles. The G-code is a collection of instructions that tells a 3D printer what motions and locations to make in order to construct a 3D object. A lamination programme named Repetier Host [27] was utilised to generate the G-code. The printing parameters were established at 225°C for the printing temperature, 0.2 mm for the layer height, 40% for the infill density, and without supporting. Depending on the 3D printer, print speed and retractions could differ. An SD card was used to upload the produced and exported G-code file to the 3D printer. The creation of the insoles was done using an MD-6C 3D printer. This 3D printer employs up to eight various thermoplastics and provides a printing size of 300 mm 200 mm 500 mm. The insoles could be manufactured even without technical difficulties because to the larger workspace. Both stretchable materials were used in the production of the two pairs of insoles.

3.14.3.1.3.1. Result

Foot scanning marked the beginning of the production process. This body part's 3D model was correctly reconfigured. The insole's 3D model was created by Gensole utilising FDM technology. In less than two days, two pairs of flexible polymer insoles were produced. The 3D printing process took the longest of all the processes in this workflow, taking over 16 hours for a pair of insoles. The insole didn't require any redesigning or further processing. Figure 3.14 depicts the outcome of the insole created with material A. Following the creation of insoles made of materials A and B, the patient put them to use by taking three walks, each with a different kind of insole.

3.14.4. CASE STUDY: QOROX 3D PRINTING TECHNOLOGY

Construction industry productivity may be raised by implementing the cutting-edge 3D printing technology developed by QOROX.

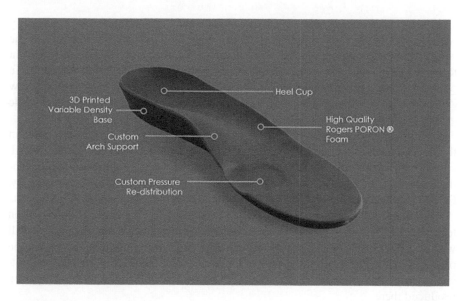

FIGURE 3.15 3D Printed Insole.

The first hybrid timber and concrete home with 3D printed concrete walls was built in New Zealand by one creative supplier, QOROX, demonstrating how 3D printing technology may be utilised to boost efficiency in the construction business. The secret to this success is perseverance, hard effort, and clients that are open to embracing new technology and collaborating to develop innovative solutions to the industry's difficulties.

In order to create an approach that is comparable to that of pre-cast wall panels, which have been widely employed in the building industry, QOROX adopted new technologies incrementally. The advantage of a 3D printing strategy is that it does away with the costly moulds used to produce pre-cast wall panels. According to the conventional formwork used for cementitious materials, 3D printed panels can have a permanent, appealing look that forms a part of the finished construction.

QOROX actively sought out forward-thinking innovators and discovered regional councils who were eager to innovate in order to test the technology on a real project. Smaller construction companies and architects are now considering alternate techniques as a result of the present market's escalating inflation and persistent supply problems with standard building supplies.

3.14.4.1 Approach

Slightly earlier on in the design phase, QOROX made a significant investment in a thorough product validation process with Building Research Association of New Zealand (BRANZ). This assured both of the customer and the Building Consent Authority that 3D printed walls could provide the same effectiveness under the New Zealand Building Code as more conventional methods.

In order to create a combination that would enable the concrete to quickly solidify while it was being printed, QOROX had to originally acquire the majority of the components from other countries when testing 3D technology. The business sought to create a concrete mix that could be made with resources found in the nation in order to create a marketable product that could be utilised on a large scale in New Zealand. This would assist to reduce shipping costs and the ecological effect while giving clients peace of mind about having a trustworthy source of supplies.

Callaghan Innovation gave QOROX vital assistance by providing cash to assist in the development of the New Zealand concrete mix and by connecting QOROX with key players in the building and construction industry. For the purpose of developing a strategy for testing, compliance, and commercial adoption of the technology, QOROX collaborated with BRANZ, Auckland University, and industry specialists.

Outcomes and benefits – speed: When compared to conventional construction techniques, 3D printing can speed up construction and cut waste significantly. A wall with a surface area of 6.5 sq m and dimensions of 2.5 m x 2.6 m x 0.25 m may be produced in around 60 minutes. Due of the mobility of the printers themselves, outcomes can either be produced at a factory and transported to the location, or even produced there.

Reduced demand for labour and training: Comparing 3D printed concrete to more conventional building techniques, the training needs are far lower. Getting certified as a brick and block layer might take two to three years. On the other hand, an operator receives complete instruction on how to utilise the printing machinery after four weeks of training. Although it is not necessary to have prior machine expertise, QOROX often hires individuals with a background in the building profession.

No specialist software for the design team: Additionally, engineers and designers do not need to learn how to utilise specialised software if they submit their drawings to be printed. They merely need to become accustomed to the new style of creation procedure as QOROX works with DWG or DXF design files.

Improved quality of construction: The 3D printer receives construction directly from the design. It is possible to obtain and verify a high level of quality prior to the start of construction. To produce a pre-finished look, more complex geometry and minute detail might be utilised.

Reduced labour and materials demand: The technology possesses the same potential to decrease the amount of on-site labour needed as other off-site strategies. It eliminates the need for certain additional tradesmen if it is used as a pre-finished item, like the Huia Bay project [16].

REFERENCES

1. www.bricsys.com/blog/8-reasons-why-students-should-try-3d-modelling
2. Slick, J. (2020). What is 3d modeling? *Lifewire.* https://www.lifewire.com/what-is-3d-modeling-2164
3. What is solid modeling? 3D CAD software. *Applications of Solid Modeling. www.brighthubengineering.com.* https://www.brighthubengineering.com/cad-autocad-reviews-tips/19623-applications-of-cad-software-what-is-solid-modeling/.
4. https://ufo3d.com/types-of-3d-modelling-different-industries

5. Yu, F., Lu, Z., Luo, H., & Wang, P. (2011, Febraury 3). *Three-dimensional model analysis and processing*. Springer Science & Business Media.
6. https://all3dp.com/2/types-of-3d-modelling/#:~:text=Within%20CAD%2C%20there%20are%20three,specialized%20for%20their%20specific%20purposes.
7. Repetier Software. www.repetier.com/
8. https://transport.itu.edu.tr/docs/librariesprovider99/dersnotlari/dersnotlarires112e/not/cadd-14.pdf?sfvrsn=4
9. www.spatial.com/resources/glossary/what-is-surface-modelling#:~:text=Surface%20Modelling%20is%20the%20method,validate%20imperfections%2C%20and%20apply%20smoothness
10. Arefin, A. M. E., Khatri, N. R., Kulkarni, N., & Egan, P. F. (2021). Polymer 3D printing review: Materials, process, and design strategies for medical applications. *Polymers*, *13*(9), 1499. https://doi.org/10.3390/polym13091499
11. Sorby, S. A. (2000). *Solid modelling with I-DEAS*. Prentice Hall.
12. Bertoline, G. R. (1997). *Technical graphics communication*. WCB McGraw-Hill.
13. Rooney, J., & Steadman, P. (1997). *Principles of computer-aided design*. UCL Press.
14. McMahon, C., & Browne, J. (1995). *CAD/CAM from principle to practice*. Addison-Wesley.
15. Tizzard, A. (1944). *An introduction to computer-aided engineering*. McGraw-Hill.
16. Right Hand Rule. (n.d.). *Georg Mischler*. www.schorsch.com/en/kbase/glossary/right-hand-rule.html
17. https://fabheads.com/blogs/what-is-the-role-of-slicing-in-3d-printing.
18. Grimm, T. (2004). *The rapid prototyping process. User's guide to rapid prototyping*. Society of Manufacturing Engineers, p. 55.
19. https://bigrep.com/posts/fdm-vs-sla-3d-printer
20. https://all3dp.com/2/3d-printer-g-code-commands-list-tutorial
21. www.stratasys.com/en/stratasysdirect/resources/articles/3d-printing-material-selection-considerations.
22. www.sharrettsplating.com/blog/materials-used-3d-printing
23. Abdulameer, H., & Tukmachi, M. (2017). Nasal prosthesis fabrication using rapid prototyping and 3D printing (a case study). *International Journal of Innovative Research in Science, Engineering and Technology*, *6*(8), 15520–15526. https://doi.org/10.15680/IJIRSET.2016.0608003.
24. TRUMPF Uses 3D Printing to Improve Satellites and Aircraft. (2019, June 17). *TRUMPF*. www.trumpf.com/en_IN/newsroom/global-press-releases/press-release-detail-page/release/trumpf-uses-3d-printing-to-improve-satellites-and-aircraft/
25 Zuñiga, J., Moscoso, M., Padilla-Huamantinco, P. G., Lazo-Porras, M., Tenorio-Mucha, J., Padilla-Huamantinco, W., & Tincopa, J. P. (2022). Development of 3D-printed orthopedic insoles for patients with diabetes and evaluation with electronic pressure sensors. *Designs*, *6*(5), 95. https://doi.org/10.3390/designs6050095
26. Construction Accord. (n.d.). *Case study: QOROX 3D printing technology*. www.constructionaccord.nz/good-practice/beacon-projects/case-study-qorox-3d-printing
27. Lueptow, R. W., Snyder, M. T., & Steger, J. (2001). *Graphics concepts with pro/engineer*. Prentice Hall.

4 Selection of Polymers for 3D Printing

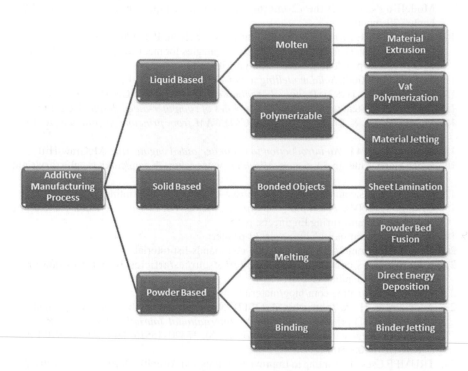

FIGURE 4.1 Different Type Additive Manufacturing Processes.

4.1. MATERIALS USED IN PHOTOPOLYMERIZATION PROCESS

Radiation-curing acrylics and acrylic blends make up the majority of the polymeric materials employed in the photopolymerization technique.

- Just where the beam hits the liquid's interface, the light-activated polymers immediately solidifies.
- After the first layer is drawn, it is dropped into the vat a short distance, and a next layer is traced on top of the initial layer.

The layers gradually link together because to the photopolymer's self-adhesive characteristic, and an entire three-dimensional structure is finally fully deposited and hardened. The patterns are then post-cured in a UV oven after being submerged in a chemical solution to eliminate any extra resin. Using such a vat with a slightly

DOI: 10.1201/9781003349341-4

flexible, transparent bottom and concentrating the UV upward through the bottom of the vat allows you to print items "bottom-up"[1, 2].

4.1.1. Materials Used in Material Jetting Process

Resins that may be photopolymerized are useful and often utilised because of their viscosity and capacity to produce droplets. However, liquid polymers may also be utilised with a printing head that has a higher temperature, and they will solidify at room temperature after printing.

This approach has the clear advantage of enabling material changes in the middle of a project. Evaluated material qualities are possible in this fashion. The following are additional aspects of the material jetting process:

- A significant subcategory is "3D printing" with low-viscosity ink.
- Although molten thermoplastics, polymer solutions, and dispersions can also be employed, light curable materials are often used.
- Wax is frequently utilised as support.

4.1.2. Materials Used in Binder Jetting Process

The binder jetting process uses two materials:

- A **powder** for part build-up and
- A **binder** to consolidate the powder

Typically, the binders is liquid. Metals, sand, and ceramics are just a few of the materials that may be printed with binder jetting [2, 3].

Sand is one of the resources that doesn't need to be processed further. Depending on the purpose and the part's required ultimate density, additional materials are frequently sintered, postured, or even occasionally penetrated with other materials.

Binder jetting is distinctive because it does not always use heat throughout the construction process. Other methods of additive manufacturing use a heating element, which might leave residual tensions in the component pieces. In a subsequent post-processing step, these pressures must be reduced.

4.1.3. Materials Used in Powder Bed Fusion

Both amorphous and crystalline thermoplastic particles can be employed as polymer powders in procedures involving powder bed fusion. Popular polymers entail: polyamide, polyamide with glass filler, polyetheretherketone, and polystyrene

The most common alternative is polyamide 12, either pure or mixed. The use of elastomeric polymers like TPE and TPU as well as polyamides 11, polyamide 6, and polyamide 11 is expanding. Epoxy and other thermosetting powders have also been employed to create pieces made entirely of plastic as well as a binder for metal or ceramic particles. An outstanding dissertation on the selective laser sintering (SLS) consolidation of polymer powders is provided by Kruth6 et al., while Schmid7 et al.

offer details on the confluence of both intrinsic and extrinsic polymer characteristics required to produce a polymer powder probable for SLS use.

The most common alternative is polyamide 12, either pure or mixed. The use of elastomeric polymers like TPE and TPU as well as polyamides 11, 6, and 11 is expanding. Epoxy and other thermosetting powders have also been employed to create pieces made entirely of plastic as well as a binder for metal or ceramic particles. An outstanding research on the selective laser sintering (SLS) consolidation of polymer powders is provided by Kruth et al., while Schmid et al. offer details on the confluence of both intrinsic and extrinsic polymer characteristics required to produce a polymer powder probable for SLS use [4].

4.1.4. Tailoring Materials for FDM

A wide range of technical thermoplastics are used in FDM, although acrylonitrile-butadiene-styrene copolymers (ABS), polylactides (PLA), polycarbonates (PC), and polyamides (PA) stand out as the most significant ones [5]. Studies have looked at how component orientation and route design affect the anisotropic mechanical characteristics of FDM-built ABS parts. To facilitate FDM processing and enhance the material qualities of printed items, many organisations have created ABS derivatives. Masood et al., for instance, looked into how metallic filler content affected rheological characteristics, ideal process settings, as well as thermo-mechanical properties. In order to assess the thermal and rheological characteristics of industrial PHAs, Kovalciket et al. [6] created scaffolds using poly(3-hydroxybutyrate) (PHB), poly (3-hydroxybutyrate-co-3-hydroxyvalerate) (PHBV), and poly(3-hydroxybutyrate-co-3-hydroxyhexanoate) (PHBH), as well as poly(lactic acid). They discovered that PHBH was the best material to use for constructing scaffolds using FDM for applications involving tissue engineering Ansari et al. created a dual-extrusion technique that outperformed FDM generated ABS by creating filaments from such a polypropylene mix comprising thermotropic liquid crystalline polymer (20 wt%) [7]

4.1.5. Polymeric Materials for Inkjet Printing

Both the building and support materials must have a low enough viscosity at the print head temperature for inkjet additive manufacturing. Urethane acrylate-based resins having viscosities between 10 MPa and 16 MPa s at temperatures between 70 and 90 °C are described by Schmidt et al. [8]. The resins contain 5–15 weight percent of an inert urethane wax and 20–45 weight percent of either tetrahydrofurfuryl methacrylate (20–45 wt%) or triethylene glycol dimethacrylate as reactive diluents. The construction material is supposed to be held partially in place by the wax, which freezes at 40 °C, until it is photocured. Resins for inkjet AM must have extremely strong heat stability and cure quickly when exposed to light in order to handle lengthy print jobs (five or more hours). Epoxy monomers, that are so frequent in AM produced by vat photopolymerization, are so rarely employed in AM produced using inkjet. For inkjet AM, the function of support material is crucial, and material development is the subject of multiple patents. Based on water-soluble monomers and polymers that photocure to produce a purposely fragile material that can be removed with water, the initial support material created by Objet (now Stratasys) was designed to be used in 3D printing. The main element of such systems, together with PEG mono- and diacrylates, photoinitiators, stabilisers, and silicone surface modifiers, is nonreactive poly(ethylene glycol, or PEG).

4.2. POLYJET

The programme uses a 3D CAD file to automatically determine the positioning of photopolymers and support material during the pre-processing. The resin 3D printer sprays small droplets of liquid photopolymer while printing, which the UV light instantaneously cures. On the build tray, tiny layers stack together to produce a number of intricate 3D printed objects. The Polyjet 3D printer fires a detachable support material called FullCore 750 where eaves or complicated geometries need support. Until the item is complete, the fine layer polymerization method is repeated. Following photopolymerization, the batch is subjected to a pressured water jet. This makes it possible to eliminate any extra fluids or supports with a little amount of individual engagement. The finished product has a surface that is inherently smooth and may be further polished to achieve nearly complete transparency for the clear resin. The advantages of both plastic and powder-based technologies are combined in Polyjet Process, an entire 3D printing technology. Furthermore, it offers the finest combination of strength, speed, quality, dependability, and flexibility in a single print. Therefore, Polyjet is the way to go if you want to give your prototype a little more in terms of materials, aesthetics, and precision. Eight different material types with distinctive properties are available from Polyjet for use in single-material printing, multi-jetting, or hybrid printing [9].

4.3. DIGITAL MATERIALS

A sort of integrated material used in Polyjet 3D printing is called digital material. You can create digital materials that combine up to six distinct resins with multi-jetting 3D printers. The models might also be as intricate and complicated as required. Consequently, enabling you to be as imaginative as possible with the results. Numerous different colour options can also be supported by these flexible or stiff materials.

4.3.1. Digital ABS Plastic

Digital ABS polymers are more heat resistant and tougher than standard ABS plastic. The very thin walls of prototypes produced of this material can have improved dimensional stability. Digital ABS plastic may be used for 3D printing of electrical components, engine components, smartphone cases, and other products that need a snap-fit assembly. Rubber-like Polyjet materials can be used to 3D print prototypes for products that require some flexible or softer coatings. Items that require non-slip surfaces, soft-touch grips and handles, valves, seals, and hoses, as well as athlete-specific footwear prototypes, are all included in this list. Additionally, a variety of Shore hardness values, from Shore A 27 to A 95, may be created by combining rubber-like materials with stiff materials.

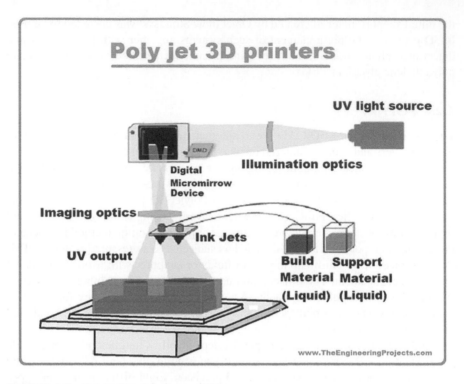

FIGURE 4.2   Polyjet Printing Process [10].

4.3.2. HIGH-TEMPERATURE MATERIALS

Prototypes and models exposed to heat benefit from outstanding dimensional stability when made using high-temperature Polyjet materials. The hardness and flexibility of their insulation can be increased by mixing high-temperature materials with rubber-like compounds. Polyjet high-temperature substances are great for working prototypes that need to tolerate a lot of heat or work under bright illumination.

4.3.3. TRANSPARENT MATERIALS

Transparent Polyjet materials are excellent for prototypes that require see-through components. The translucent materials may also be coloured, giving you an astounding array of colours for your figure.

Such materials are perfect for creating prototypes for glass items, eyeglasses, product coverings, and a variety of medical devices including contact lenses and transparent dental appliances.

4.3.4. RIGID OPAQUE MATERIALS

Polyjet stiff transparent materials are a good option to consider if you want the greatest variety of colour possibilities for the design. They may be used to make moving or assembled pieces as well as electrical components when mixed with rubber-like materials. They also produce things that need to be pleasant to the touch and colourful to be easily recognised.

4.3.5. SIMULATED POLYPROPYLENE MATERIALS

Materials made with simulated polypropylene mimic the properties and look of the real stuff. For instance, this kind is used in a wide range of applications, including product packaging, lab equipment, automotive parts, living hinges, and music speakers. Before moving into full manufacturing, create a polypropylene prototype that can be tested or displayed using 3D replicated polypropylene.

4.3.6. BIOCOMPATIBLE MATERIALS

The five essential criteria for medical application are met by Polyjet biomaterials. This factor consists of USP Class VI plastic, delayed hypersensitivity, irritation, genotoxicity, and cytotoxicity. These materials can be used to make colourless prototypes that need brief contact with mucosal membranes for at least 24 hours. Additionally, prototypes that will come into contact with the skin for a longer period of time than one month can be created.

4.4. ELECTRON BEAM MELTING

The majority of research is in the area of alloys because there are few uses for pure metals. A restricted number of metals, such as titanium alloys (perfect for medical implants), cobalt chrome, steel powders, and nickel alloy 718, can be utilised with EBM. These materials have excellent mechanical qualities, high strength, and corrosion resistance, all of which are very beneficial in demanding applications. Any material utilised in EBM must be conductive because the method significantly depends on electrical charges. Pure titanium for use in medical applications, pure copper because of its exceptionally high thermal and electrical conductivity, and Nb because of its highly conductive qualities are the only outliers. Pre-alloyed powders are often utilised. However, in situ alloying utilising powder mixtures and in-situ reaction have both been studied. The creation of cellular materials with specific characteristics is another intriguing SEBM application area. In general, if the right process conditions are employed, metals and alloys may achieve about 100% density. It is crucial to distinguish between hot isostatic pressed (HIP) (+ heat treated) materials and materials that have been heat treated as well as those that have been constructed. Additionally, texturing may result in features that rely on their orientation. There is a significant difference between non-machined and machined samples due to the high surface quality. The position within the build chamber has a significant impact on the final characteristics, particularly for the as-built materials.

Cellular Material: In addition to enabling the creation of intricate, dense components, additive manufacturing may also be utilised to create open-cellular materials with a significant level of porosity. The cell layout of cellular materials 88–90 determines their unique characteristics, that can only be somewhat changed by employing traditional powder metallurgy or melt manufacturing methods. Unlike other tech-niques, AM enables the creation of almost any open-cellular designs from high-performance materials. The cells architecture is developed by the deterministic printing process rather than a stochastic self-organization process, like foaming. This design flexibility is crucial because it enables the use of EBM to create and realise cellular materials with exceptional features [1, 11].

4.5.   SUPPORT MATERIAL

Whenever a print contains overhangs or elements hovering in midair, support struc-tures are required for FDM 3D printing. They stabilize these normally unsupported sections, enabling effective printing of complicated forms.

The substance on which these supports are printed is known as 3D printer support material. Choosing the right material for creative models will make your printing experience lot more joyful since different materials provide varied balances between price, convenience of usage, and printability [12].

4.5.1.   BASIC: BUILD MATERIAL SUPPORTS

The very same material that your model is constructed of is the easiest, most widely used support material. This is due to the fact that many 3D printers are still single-extruder devices that can only print particular material at a moment, which makes it hard to employ a specific support material.

Common building materials are often more readily available and more cheap than specialised support supplies. As a result, building material supports could be a desir-able choice for individuals with a tight budget [12].

Because they are constructed using the same substance as your model, build material supports are also certain to stick to your prototype better. However, this stickiness has two disadvantages: Prints are less probable to fail, but removing the support will take more work and provide a lower-quality surface. As was seen previously, to get a smooth texture on the model following support elimination, an X-Acto knife or sandpaper could be required.

Build material supports are the best option if you want to ensure adhesion, save some money, and also don't mind laborious support withdrawal and yield loss. And, for greater or even worse, this is the only choice you have if you own a single-extruder 3D printer.

Works on: Single-extruder, multi-extruder 3D printers
Pros:

• Works on single-extruder 3D printers
• Often more economical
• Material compatibility is not a concern

Cons:

• Poorer surface quality
• Support removal can be a hassle

4.5.2. Quick Removal: Breakaway Supports

Like building material supports, loose supports function similarly. They are far less likely to overly cling to the prints nevertheless because they are made of a different substance. The identical support networks will thus be considerably simpler to dismantle and leaving behind cleaner surfaces if they are printed on breakaway materials.

However, breakaway supports are not flawless. As is frequently the case with multi-material 3D printing, material compatibility is a worry since the support material could not attach effectively to all construction materials. The least popular choice on this list, breakaway support materials are also not readily available. Alternatives include the PRO Series Breakaway from Matterhacker, the Breakaway from Ultimaker, and the Scaffold (also water-soluble) and Scaffold Snap (only breakaway) materials from E3D.

A specific breakaway support material is what you need if you want a support that easily rips off and retains a smooth surface finish.

Works on: Multi-extruder 3D printers
Pros:

- Quick and clean removal

Cons:

- Material compatibility is a consideration
- Poor availability
- Only works on multi-extruder 3D printers

4.5.3. Best Quality: Soluble Supports

The superior surface quality is provided by soluble support materials. Soluble supports don't need to be manually removed; they just disappear, leaving completely clean, unblemished surfaces. This enables complicated geometries that solvents can penetrate but pliers or X-Acto blades cannot, as well as thick, solid supports for maximum dimensional precision. Such thick supports and clean removal are generally advantageous for moving components, for example the previous gyro. But there is a cost involved. Not only are soluble support materials figuratively more costly, but they are also sometimes more challenging to handle. PVA, the most popular water-soluble substance, is particularly hygroscopic, meaning that moisture causes it to break down. It's crucial to store filaments properly. Compound compatibility is a problem as well since no one soluble material is compatible with all types of construction materials. Alternatives like HIPS, which are less than ideal if PVA doesn't work, sometimes require unpleasant chemicals to disintegrate. Furthermore, soluble supports may take up to hours to completely dissolve, which makes them a bad option for people who are pushed for time.

Soluble support materials are really for you if you want the greatest maximum performance, no matter how you store your filament, how long it takes to dissolve, or even if you want to avoid potentially dangerous chemicals.

Works on: Multi-extruder 3D printers
Pros:

- Cleanest removal and surface finish
- Complex structures can be supported (internal structures, moving parts)

Cons:

- Material compatibility is a consideration
- Material storage can be a hassle
- Dissolving times are long
- Only works on multi-extruder 3D printers

4.5.4. NOT ALL OVERHANGS NEED SUPPORTS—THE 45 DEGREES RULE

Not every overhang requires support. The basic rule of thumb is that you might be able to create an overhanging without utilising support networks for 3D printing if something tilts at an angle only around 45 degrees from the perpendicular.

A very slight horizontally offset (barely perceptible) is used by 3D printers among successive layers. A layer therefore stacks with a little offset rather than exactly over the one underneath it. As a result, overhangs that don't tilt too much from the vertical can be printed by the printer. The preceding layers can sustain anything that is north of 45 degrees, but anything south of that angle cannot. According to others, the breaking line is 45 degrees.

The letters Y and T are the best examples for illuminating this aspect. The angle between the vertical and the two overhangs in the letter Y is less than 45 degrees. Because of this, printing the letter Y is possible without the use of any 3D printing supports.

In contrast, the awnings in the letter T form a 90 ° angle with the vertical. In order to avoid a mess like the one shown in Figure 4.3, formwork for 3D printing must be used to print the letter T [13].

4.5.5. BREAKAWAY 3D PRINTING SUPPORT STRUCTURES

Breakaway 3D printing support structures are used by default in printers that have a single extruder.

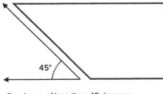

Overhang of less than 45 degrees
No support is needed

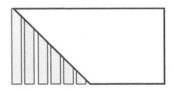

Overhang of more than 45 degrees
Support is needed

FIGURE 4.3   Support Structure 45° [14].

For the purpose of creating the 3D printing support networks when you only have one extruder, you must utilise the same material that was used to print the design. The first and only control you have over support material is the ability to gradually change the density of the 3D printing support structures to make it dramatically decrease than the design densities.

The only method to remove the model from the 3D printing support structures is to gently tear the support structure off with a knife or break it off using a hammer.

These removal techniques carry a significant danger of injuring the model, thus it is imperative to use good technique and exercise utmost caution and vigilance throughout the separation process.

4.5.6. DISSOLVABLE 3D PRINTING SUPPORT STRUCTURES

There is a better solution if your printer has two extruders. One extruder can be loaded with PLA to print the model, while another can be loaded with a substance that dissolves in water or limonene, such as PVA or HIPS, to print the support structure in place. After printing is finished, just submerge the model in water or limonene to remove the support structure.

Complex prints are best suited for this technique of removal since it lowers the possibility of model breakage and enables post-processing operations simpler.

When the first unicellular organisms on Earth initially formed, polymers like proteins and nucleic acids were already there. For thousands of years, people have used other natural polymers like cellulose and starches for things like food, clothing, and shelter. In the 19th century, cellulose, polyisoprene, and shellac were transformed into practical man-made polymers, fibres, and elastomers, although these transformations were mostly guided by empirical observation. The degree of uniformity in the polymeric matrix and the overall strength of the intermolecular secondary valence bonds are major determinants of the variations between fibres, plastics, and elastomers. The latter might be London forces, relatively powerful dipole–dipole forces, or strong intermolecular hydrogen bonds.

The amount of repeating units, or mers, inside the polymer chain determines the distance of the polymeric chains in HDPE as well as other polymers; this number is denoted by the letter n or the acronym DP (degree of polymerization). Of course, M's molecular weight, or molar mass, is equal to n times the molecular weight of the repeating unit, m, or $M = nm$. Therefore, M will equal 2800 if n for HOPE-$(CH2CHdn$ is equal to 1000. The interactions between the polymeric chains are not as strong as they may be because of the moderate London forces that operate between molecules. The word "oligomer" refers to relatively low molecular weight polymers with n values less than 20 because it comes from the Greek word oligo, which means "few." Because of their alleged unrestricted rotation, the carbon–carbon (C C) bonds in HOPE are adaptable. At temperatures much above glass transition temperature Tg—defined as the point at which this segmental motion starts when the temperature of the polymer is raised—the polymer chains are in a continual state of so-called rotational movement, also known as a wiggling segmental motion. In general, the molecular weight and structural regularity of a polymer, like polyethylene, rise with increasing molecular weight, whereas the amount of random

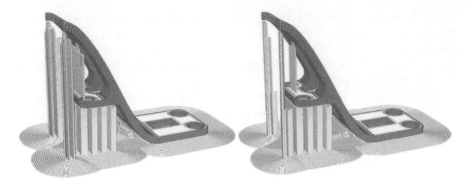

FIGURE 4.4   Hanging Support Structure [15].

branching in the polymer molecule decreases with increasing molecular weight. Hope crystallises easily like linear paraffins due to its uniform structure. As a result of its more erratic structure, branching or low-density polyethylene (LDPE) is a little less crystalline.

4.6.   PHYSICAL STRUCTURE OF POLYMERS

Four physical states of polymers are possible: crystalline, three amorphous, and one intermediate form (glassy, rubbery, and viscous flow). Rigid polymers are also referred to as polymers that exist in the glassy or crystalline state. Each condition has a unique set of mechanical characteristics and a specific field in which they can be used technically. The most common method for determining a polymer's physical state is to examine its mechanical characteristics, such as compliance. The fundamental characteristic of chain molecules is flexibility. A fibre with a diameter of 10 microns and a length of several millimetres may be made from polymer molecules, which have a transverse dimension of several angstroms and a length of several thousand angstroms [11].

4.6.1.   MELT VISCOSITY

It is significant to note because polymer chain entangled below a certain crucial threshold molecular weight is minimal at best. A measurement of the propensity (speed) of melted materials to move is called melt viscosity [16]. A polymer's melt viscosity rises in direct proportion to molecular weight up to a crucial threshold molecular weight as it gains weight. The melt factor, which measures flow in an inverse relationship to melt viscosity, is frequently used in specific terms, such as the amount of time it takes for ten grammes of a material to transit through some kind of typical orifice at a certain temperature and time. The melt viscosity is correlated with the molecular weight increased to the 3.4th power just above that number. For polymethyl methacrylate (PMMA), polyisobutylene (PIB), and polystyrene (PS), the essential threshold chain lengths are 208, 610, and 730 repetitive units, accordingly.

4.6.2. INTRA- AND INTER-POLYMER CHARACTERISTICS

Primary forces and auxiliary forces are frequently used to categorise the forces found in molecules. Foremost bonding forces can also be divided into three categories: ionic (between atoms of greatly different electronegativities); metallic (the number of outer, number of electrons in valance band is just too small to complete outer shells; metallically bonded atoms are frequently thought of as charged atoms surrounded by a potentially fluid sea of electrons; lack of bonding direction; not found in polymers); and coval (between atoms of similar electronegativities). The distance of a primary bond is typically between 0.09 and 0.20 nm, whereas the length of a carbon-carbon bond is between 0.15 and 0.16 nm.

The van der Waals modification to the ideal gas equations is caused by auxiliary forces, also known as van der Waals forces, which interact over greater distances than main forces, often having considerable interaction between .25 and .50 nm. Since both impact how close together chains are in a polymer, many physical features of polymers rely on both the morphology (arrangements linked to rotate about single bonds) and the geometry (arrangements related to the actual chemical bonding around a specific atom).

4.6.3. INTRA-POLYMER STRUCTURE CHARACTERISTICS

Polymer chains are often "soft," "stiff," or somewhere in between. A long chain polymer's "stiffness" or "softness"—or anything in between—depends on the intra-polymer structural properties of the polymers.

4.7. INTER-POLYMER FORCES

Between its chains, certain polymers have moderate interactions whereas others have high forces. This inter-polymer force is determined by van der Waals forces. These two elements may be used to comprehend the many characteristics of polymers as well as the reasons why they differ significantly from materials such as metals and ceramic materials.

4.7.1. GLASS TRANSITION TEMPERATURE (TG) AND ITS PHYSICAL MEANING

At normal temperatures, amorphous flexible polymer chains are often constantly in motion due to the kinetic energy stored in the molecule. As the temperature is dropped in this reversible process, the degree of this segmental motion that resembles wriggling reduces. The glass transition temperature, or abbreviated Tg, is the temperature at which the segmental or micro-Brownian motion of amorphous polymers becomes important as the temperature rises. The entire volume inhabited by the holes is referred to as the "free volume." Under the Tg, there is no change in a polymer's free volume or vacant volume when temperature is raised; nevertheless, the rate of change rapidly rises at the Tg. In fact, one way to determine the Tg is to measure the change in the slope of the curve showing free volume as a function of temperature. The unbound volume can be up to 10% of the total volume at the

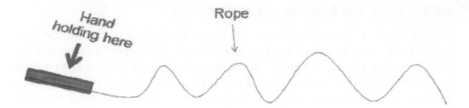

FIGURE 4.5   Quick Changes of Inter-Polymer Properties.

Tg-temperature and is mostly independent on the kind of polymer. Tg is a crucial variable that is specific to polymers. According to intra- and inter-polymer properties, the length of a polymer chain segment fluctuates. The quick changes in form that occur when a rope is dragged up and down are seen in Figure 4.5.

From this example, it can be shown that a polymer chain's stiffness rises with the lengths of each segment. Tg is the temperature at which polymer segments start to flow out of their frozen condition (with a rise in temperature) or begin to freeze (with a drop in temperature). This modification to the chain section is shown in Figure 4.5.

4.7.2.   Melting and Crystallization Temperatures Tm and Tc

Polymers do not always crystallise, in opposed to the bulk of inorganic crystal structures. Amorphous and crystalline phases are always present in polymers. The crystallinity of the substance, not its Tg, decides whether it is a plastic or a rubber. The PE behaves more like a strong plastic than a rubber because of the crystalline (hard phases) and rubbery (soft phases) components. Re-crystallization occurs during heating before melting, which makes polymers more complicated.

4.7.3.   Crystallinity

The inclination of a particular polymer to just be crystalline, amorphous, or to have a combination of crystalline and amorphous areas in the solid phase depends on its chemical structure. Symmetrical chain architectures [17], which enable compact packing of the polymer units to better capitalise on secondary forces that are linked to distance, are generally favourable for crystallinity. High interchain (secondary forces) interactions also encourage crystallinity. As a result, the extremely symmetrical chain architectures of PTFE and linear polyethylene (HDPE) give them a strong propensity to crystallise. With an asymmetrical chlorine atom, atactic PVC has a propensity towards amorphousness. But since two chlorine atoms in polyvinylidene chloride (PVDC) are symmetrical, PVDC has a tendency to be crystallised. Because of certain interchain interactions, in this example h—bonding, atactic PV A is partly crystalline.

Increases in numerous characteristics, including Tg and Tm, are often brought on by factors that boost hydrogen bonding. HDPE, although having a high degree of symmetry, has a low Tm (135°C), whereas nylon 66, which is symmetric and has strong hydrogen bonds, has a Tm of roughly 270°C. When extended or cold pulled at temperatures underneath the Tm, crystallisable polymers often produce randomly distributed crystallites that are orientated. Pressure-induced crystallisation can produce a fibrillar structure or an immediate and longer-term structure.

In comparison to amorphous polymers, crystalline polymers are less soluble below the Tm. Spherulites are collections of disoriented crystallites that can develop when a crystallisable polymer is cooled below Tm. A polymer's density and transparency are inversely correlated with its degree of crystallinity.

The polymer's tactility has an impact on crystallinity as well. The possibility of a macromolecule crystallising increases with increasing molecular order. For instance, isotactic polypropylene (iPP) tends to be more crystalline over syndiotactic polypropylene, although atactic polypropylene is thought to be impossible to crystallise due to the irregular structure of the polymer chain. The majority of atactic polymers, in fact, don't crystallise.

In comparison to amorphous polymers, crystalline polymers are less soluble below the Tm. Spherulites are collections of disorganized crystallites that can develop when a crystallisable polymer is cooled below Tm. A polymer's density and transparency are inversely correlated with its degree of crystallinity.

The polymer's tactility has an impact on crystallinity as well. The possibility of a macromolecule crystallising increases with increasing molecular order. For instance, isotactic polypropylene (iPP) tends to be more crystalline over syndiotactic polypropylene, although atactic polypropylene is thought to be impossible to crystallise due to the irregular structure of the polymer chain. The majority of atactic polymers, in fact, don't crystallise.

X-ray crystallography may be used to evaluate crystallinity, although calorimetric methods are also often employed.

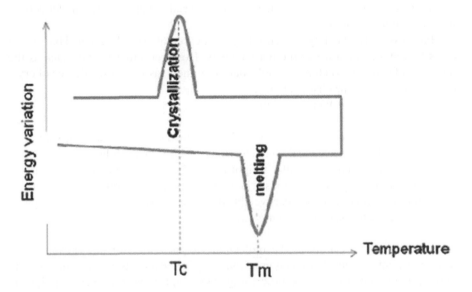

FIGURE 4.6   Different Scanning Calorimetry.

Around the maximum temperatures Tm and Tc, both melting and crystallisation occur across a defined range of temperatures. The re-crystallization upon heating is caused by the unfinished crystallisation process. Re-crystallisation will thus occur when the polymer is heated, albeit at temperatures lower than the melting point of the present crystals. Therefore, Differential Scanning Calorimetry cannot be trusted to detect polymer crystallinity.

For crystalline polymers, the relationship between T_m and T_g has been described by Boyer as follows

$$T_{g°} \, / \, °T_m \approx °1 \, / \, 2 \rightarrow \text{symmetrical polymers} \tag{1}$$
$$T_{g°} \, / \, °T_m \approx °2 \, / \, 3 \rightarrow \text{unsymmetrical polymers} \tag{2}$$

It was discovered that many polymers with a Tg/Tm ratio less than half are extremely regular and have a small number of repeating units including one or two main-chain atoms each containing just one substituent. Chemical composition and thermal history, including such cooling conditions during production or post-thermal treatment, have an impact on the degree of crystallisation. The observed melting enthalpy, ΔH_{meas}, is placed in reference to the literature value, ΔH_{lit}, for totally crystalline material in order to determine the degree of crystallisation, K.

$$K = \Delta H_{meas} \, ° \, / \Delta H_{lit} \tag{3}$$

4.7.4. THERMAL HISTORY

The first heating curve of a DSC test displays the thermal or mechanical history. The second heating curve is used to calculate the material's characteristics under certain dynamic circumstances.

Hardness is significantly influenced by the degree of crystallisation, The ratio of mass to volume is known as the mass density. However, other factors, such as the diameter of cellular structures or molecular alignment, as well as the degree of crystallinity, can affect the characteristics.

Polymer Molecular Weight and its Meanings:

As the molecular weight rises to the threshold molecular weight, several physical and mechanical characteristics of amorphous polymers progress quickly. To show molecular weight as a polymer characteristic, there are two commonly used methods. The first is the weight average Mw, whereas the second is the molecular weight Mw average. A true molecular weight distribution is shown in Figure 4.7.

The viscosity rises sharply as the molecular weight rises over the threshold molecular weight. An ideal or economic range is frequently used for commercially multi purpose polymers since processing these high-molecular-weight polymers

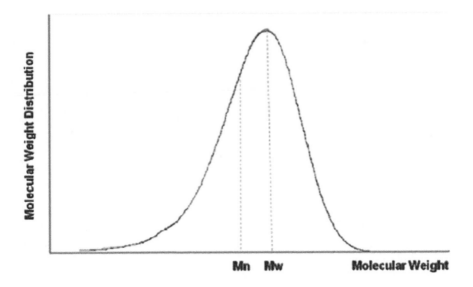

FIGURE 4.7   Molecular Weight Distribution.

uses more energy. Because there are far more chain ends in each unit of volume in lesser molecular weight polymers, those with molecular weights over the entanglement threshold are more flexible than higher-molecular-weight counterparts. The packing effectiveness of the smaller molecule weight polymers is decreased by the chain ends.

The packing effectiveness of the lower molecular weight polymers is decreased by the chain ends. Some significantly greater-molecular-weight polymers, such ultrahigh-molecular-weight polyethylene (UHMWPE), are employed despite their processing challenges because to their better toughness.

4.7.4.1.   Mechanical Properties of Polymers

The ability of a polymer to endure applied loads and the resulting strains over the course of its useful service life is crucial when using it as a structural element. With the characteristics of both solids and viscous liquids, polymers are viscoelastic materials. These characteristics vary with both time and temperature.

Figure 4.7 illustrates two extreme instances, a highly brittle polymer and a very ductile polymer, to show how polymers actually operate in practise.

The elevation of the earliest linear elastic area of the stress/strain curve may be used to calculate the Young's modulus (E) of the polymer. The curve may be used to determine many mechanical characteristics, such as:

- Initial yield stress prior to the polymer yield.
- Minimum stress after initial yield.

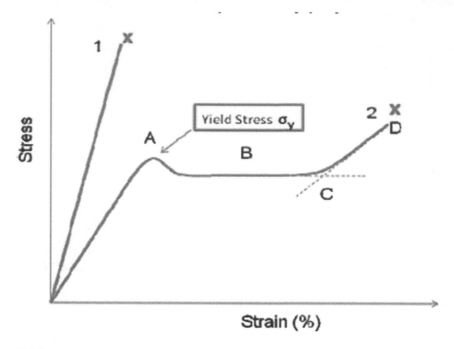

FIGURE 4.8   Tensile Stress vs Strain.

- Strain hardening, when the gradient starts to rise as a result of the polymeric chain being fully extended in the direction of the stress (higher stress is needed for further deformation).
- Failure stress at which the polymer has attained its maximum allowable stress.

4.7.5.   CRYSTALLINE EFFECTS THROUGH YIELD

Increases in crystallinity also result in increases in elastic modulus, rigidity, yield stress, and durability. Slower cooling and annealing, two physical processes that improve crystallinity, also tend to enhance a polymeric material's stiffness, hardness, and modulus. As a result, crystallinity-containing polymers are denser, stiffer, and stronger compared to amorphous polymers. However, the amorphous area helps to make polymers strong and flexible. It is possible to apply the early studies of viscoelasticity conducted on silk, glass, and rubber to fibres, plastics, and elastomers, respectively. One of the earliest findings was that certain materials, when stressed over an extended period of time, have a very gradual, irreversible flow or creep. Flat regions are caused by this irreversible phenomenon in nylon 66-reinforced pneumatic tyres.

When polymers are exposed to cyclic loading, they may potentially fail. These fatigue data are often shown against the stress concentration is S in the specimen through using log of the number of alternating loads that cause failure, N. The stress under which no failure happens irrespective of the number of cycles is known as the endurance strength or limit.

4.7.6. Viscosity

4.7.6.1. Elastomers

We use the word "viscosity" to refer to the "thickness" of various liquids. We may remark, for instance, that honey has greater viscosity than water. Gasoline is less viscous than motor oil. That is to say, water is a lot simpler to mix or pour than honey. More slowly-moving than the honey. When we swirl it, it opposes our movements with the spoon. "Resistance to flow" is a fairly wide word that is frequently used to define viscosity. When compared to a substance like The honey doesn't flow very easily, like water.

Honey is not a polymer right now. The solution is quite concentrated. It includes various tiny molecules made by the plants from which the bees collected the nectar to manufacture the honey, as well as a small amount of water and a lot of carbohydrates. Also not polymers, the sugars are straightforward monosaccharides like glucose and fructose. However, honey shares this high viscosity not just with certain oligomers, which are short-chain polymers that may be fluids instead of solids, but also with polymer solutions. What makes honey so thick? In part, this is due to the fact that the water molecules in the presence of sugar feel far higher drag than water molecules do while moving through with a fluid.

Similar principles apply in the motor oil illustration. Hydrocarbons, which are made of carbon chains coated with hydrogen atoms, are the same family of molecules that make up motor oil and gasoline. They resemble one another a much. The intermolecular interactions between the molecules are just London dispersion forces and substantially weaker than those between sugar molecules. The longer hydrocarbon chains in motor oil make it significantly different from gasoline in terms of composition. Because the molecules are larger, they travel through the liquid with increased drag. A significant component in determining how firmly two molecules adhere to one another is the quantity of intermolecular contacts. This aspect is particularly significant in the case of extremely weak London dispersion forces, when a slight advantage can have a significant impact. Gasoline flows considerably more freely than motor oil because the larger hydrocarbon chains in engine oils stick to one another much more firmly than the shorter ones. If oligomers as well as polymers are liquids, they both exhibit high viscosity for the same reasons. Their lengthy molecule chains' prolonged interaction increases the intermolecular attractions that result in flow resistance. Enhanced drag is another contributing aspect; even in solution, these very big molecules meet greater resistance when they pass solvent molecules than smaller ones would. In actuality, measuring the viscosity of polymer solutions is another method for figuring out the size of the polymer, which then yields the chain length and molecular weight. The drag and intermolecular interaction increase with polymer size, increasing viscosity as a result.

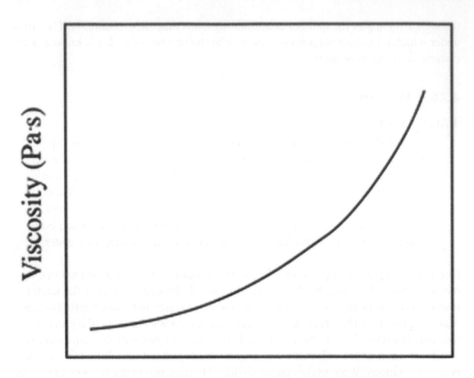

FIGURE 4.9   Polymer Viscosity.

As a result, although probably not necessarily linearly, viscosity increases with molecular weight. This may make it more difficult to use a graph like this to deduce molecular weight from the measurement of viscosity. In this situation, taking the numbers' log is a usual course of action. That method typically produces a horizontal line when two things are joined.

Elastomers are high-molecular-weight polymers that have chemical or physical cross-linking. The elastomer needs to be amorphous in its unextended, natural state, above the glass transition temperature, and used in industrial applications at a temperature that allows for full "chain" mobility. The restoring force is mostly entropic after elongation. As the material gets longer, the random chains are forced to fill more organised spaces. Once the applied force is withdrawn, the chains frequently return to a more erratic state. There should not be much chain movement overall. Cohesive energy interactions in elastomer chains enable rapid and straightforward growth. When stretched, an elastic chain has a high tensile strength, but a modest extension yields a low modulus. Weak cross-link density polymers frequently meet the necessary requirements. The material restores its previous shape after deformation because of the cross-linking. Elastic memory is another name for this characteristic.

4.7.6.2. Fibres

Elevated modulus and tensile strength are two qualities that define fibres. Both of these characteristics—high cohesive energies between chains and a lot of molecular symmetry—are connected to a rather high level of polymer crystallinity. Because they are typically linear and pulled (directed) in one direction, fibres have strong mechanical qualities in that direction (such as tensile strength, modulus, and flexural strength). In fibres, branching and cross-linking are undesirable because they obstruct crystal growth. However, if added to the polymer after the material has been properly drawn and treated, a tiny amount of cross-linking could boost some physical characteristics like tensile strength [11, 18].

4.7.6.3. Plastics

Plastics refers to a class of materials whose characteristics fall in between those of fibres and elastomers. As a result, plastics have variable degrees of crystallinity as well as some degree of flexibility and toughness. The prerequisite molecules for a plastic are:

- if it is linear or branched, with little or no cross-linking, it is below its Tg'
- if it is amorphous and/or crystalline, it is used below its Tm,
- if it is cross-linked, the cross-linking is sufficient to severely restrict molecular motion.

4.7.6.4. Polymer Blend

Mixture of two or more polymers (the second polymer added will be of significant quantity, i.e., > 2% by weight).
Some distinctions for clarity:

- Copolymer (Cop): Polymer formed by two or more types of- repeating units or comonomers.
- Polymer Compound: Mixture of polymer and additives/ingredients (from any material class).
- Polymer Composite: Any combination of polymer and filler in any form (from any material class).
- Solid state Material classification: Polymers, Metal, Ceramics.
- Polymer Crosslinking/curing/vulcanization: process of getting three-dimensional networks of polymeric chains (by chemical bonds).
- Physical cross links results out of localized crystallization/interactions due to secondary forces/entanglements.
- Interpenetrating Polymeric Networks (IPNs) can be of both types (i.e. resulting from physical/chemical crosslinks).
- Thermoplastic Elastomers (TPEs)—polymeric materials behaving like rubbers and having processibility of thermoplastics (could be achieved via blends/cop/chemical or physical cross linking).
- Polyelectrolytes/ionomers/ionic polymers: polymer backbone with one kind of charge (an/cation) and pendant/side groups with another kind of charge.

- Miscible—Homogeneous mixture in any proportion (will be homogeneous at molecular level).
- Soluble: Homogeneous mixture in limited proportion (up to saturation level).
- Homogeneous: just means same, and can be wrt morphology/orientation/distribution.
- Compatible: to work harmoniously (so they can be heterogeneous or immiscible but yet perform harmoniously without losing original properties).
- State: usually three states of matter—solid, liquid, and gas (just based on molecular packing at physical level).
- Phase: sometimes used as synonymous to state, but here the phase difference arises due to chemical and/or physical differences, any material or combination of materials exhibit distinct phases. For example in the liquid state mixture of water and oil, we can observe two phases. Semicrystalline polymers show phases of crystalline region and amorphous region. Immiscible polymers shows phase separation.
- Compatibilization: Incompatible/Immiscible blend + compatibilizer to get alloy (this can still possess phases).
- Interphase: a third phase (of about 2–60nm) between the two phases of immiscible polymer blend (or binary system).
- Binary = 2, ternary = 3.
- Toughened polymer: brittle polymer with increased toughness [mixing with rubber (for low temp. application)/with engineering plastics (for high temp. application)/with little plasticizer/process aid/toughener (is a compounding approach)].
- Need for blending (the driving force): instead of carrying out research in developing new monomers/polymers/copolymers with required set of properties as per requirement, and to establish it across the market, which consumes lot of time, money, and energy, it is worth developing a blend with existing materials with known properties.
- Methods of blending: Solutions mixing/melt mixing/mixing above Tg/mixing one polymer in another monomer and polymerising/mixing two monomers and polymerising simultaneously/latex mixing/powder mixing and processing.
- Mechanical mixing: mechanical mixing of polymers includes such methods as roll milling and melt mixing. In roll milling, the mixing of polymers can be accomplished by squeezing the stock between the rolls. In melt mixing the polymers are mixed in the molten state, e.g., in an extruder.
- Dissolution in co-solvent (followed by film casting, freeze/spray drying)
- Latex blending: Emulsion polymerization is employed for the preparation of rubber toughened plastic blends. The polymers should be in the latex or emulsion form. The mixing process of these micro-sized latexes and the subsequent removal of water produce excellent dispersion.
- Fine powder mixing: In this method components are taken in powder form and mixed. Followed by suitable processing technique.
- Use of monomers as solvents for another blend component, followed by polymerization as in interpenetrating networks [18, 19].

Enthalpy of mixing: The expression for this is evolved from basic relationships and is given by,

$$\Delta G = \Delta H - T\Delta S = \Delta U + p\Delta V - T\Delta S \tag{4}$$

$$\Delta F = \Delta U - T\Delta S \tag{5}$$

Flory-Huggins theory for developing an expression for free energy of binary polymer mixture: [independently derived by these two scientists] leads to key equation nos.

$$\Delta G_m = RTV[(\rho_1\Phi_1 / M_1)\ln\Phi_1 + (\rho_2\Phi_2 / M_2)\ln\Phi_2] + B_{12}\Phi_1\Phi_2 V \tag{6}$$

$$\Delta G_m = RTV\{[\Phi_1\Phi_2\chi_{1,2} / V_r] + [\Phi_1.\ln\Phi_1 / V_1] + [\Phi_2.\ln\Phi_2 / V_2]\}.... \tag{7}$$

Where, M = molecular weight of component ρ = density of component B = RTχ12/Vr, i.e., a binary interaction density parameter [18, 19].

REFERENCES

1. Körner, C. (2016). Additive manufacturing of metallic components by selective electron beam melting—a review. *International Materials Reviews, 61*(5), 361–377. https://doi.org/10.1080/09506608.2016.1176289
2. Pei, E., Kabir, I. R., Godec, D., Gonzalez-Gutierrez, J., & Nordin, A. (2021). Functionally graded additive manufacturing. *Additive Manufacturing with Functionalized Nanomaterials*, 35–54. https://doi.org/10.1016/b978-0-12-823152-4.00006-5
3. Crivello, J. V., & Reichmanis, E. (2013). Photopolymer materials and processes for advanced technologies. *Chemistry of Materials, 26*(1), 533–548. https://doi.org/10.1021/cm402262g
4. Powder Bed Fusion | Additive Manufacturing Research Group | Loughborough University. (n.d.). www.lboro.ac.uk/research/amrg/about/the7categoriesofadditivemanufacturing/powderbedfusion/
5. Ferrándiz, S., Miron, V., Juarez, D., & Mengual-Recuerda, A. (2017). Manufacturing and characterization of 3D printer filament using tailoring materials. *Procedia Manufacturing, 13*, 888–894. https://doi.org/10.1016/j.promfg.2017.09.151.
6. Kovalcik, A., Sangroniz, L., Kalina, M., Skopalova, K., Humpolíček, P., Omastova, M., Mundigler, N., & Müller, A. J. (2020). Properties of scaffolds prepared by fused deposition modelling of poly(hydroxyalkanoates). *International Journal of Biological Macromolecules, 161*, 364–376. https://doi.org/10.1016/j.ijbiomac.2020.06.022.
7. Ansari, M., Redmann, A., Osswald, T., Bortner, M., & Baird, D. (2019). Application of thermotropic liquid crystalline polymer reinforced acrylonitrile butadiene styrene in fused filament fabrication. *Additive Manufacturing, 29*, 100813. https://doi.org/10.1016/j.addma.2019.100813.
8. Schemidt, K. A., Doan, V. A., Xu, P., Stockwell, J. S., & Holden, S. K. (2008, May 27). (3D Systems, Inc.) Ultra-violet light curable hot melt composition. U.S. Patent 7378460B2.
9. Attention Required! | Cloudflare. (n,d,). https://all3dp.com/2/polyjet-3d-printing-technologies-simply-explained/
10. Davis, R. (2021, August 18). What is 3D printing? Definition, technology and applications. *The Engineering Projects*. www.theengineeringprojects.com/2021/06/what-is-3d-printing-definition-technology-and-applications.html
11. Chapter 1 Structure and Physical Properties of Polymers. (1981). *Tribology Series*, 1–29. https://doi.org/10.1016/s0167-8922(08)70735-1

12. Amfg. (2020). 3D printing support structures: A complete guide. *AMFG*. https://amfg.
    ai/2018/10/17/3d-printing-support-structures-guide/

13. Kapoor, A. (2021, September 3). 3D printing support structures—Makenica 3D printing.
    Online 3D printing, vacuum casting, injection moulding, CNC machining. *Makenica.
    com*. https://makenica.com/3d-printing-support-structures/

14. Manufactur3d. (2020, July 30). Designing for FDM 3d printing: 9 important tips –
    manufactur3d. *Manufactur3D*. https://manufactur3dmag.com/designing-for-fdm-3d-
    printing-9-important-tips/

15. Cura Support Settings, from Angles to Z Distance. (2021, August 1). *Wevolver*. www.
    wevolver.com/article/cura-support-settings-from-angles-to-z-distance

16. Frankland, J. (2020, September 10). Understanding the effect of polymer viscosity on
    melt temperature. *Plastics Technology*. www.ptonline.com/blog/post/understanding-the-
    effect-of-polymer-viscosity-on-melt-temperature

17. Libretexts. (2021, September 13). 4.5: Crystallinity in polymers. *Chemistry LibreTexts*.
    https://chem.libretexts.org/Bookshelves/Organic_Chemistry/Book:_Polymer_
    Chemistry_(Schaller)/04:_Polymer_Properties/4.05:_Crystallinity_in_Polymers

18. National Research Council. (1977). *Structure-property relations in polymers. Organic
    polymer characterization: Report*. The National Academies Press. https://doi.org/10.
    17226/20350.

19. National Academies of Sciences, Engineering, and Medicine. (1977). *Organic polymer
    characterization: Report*. The National Academies Press. https://doi.org/10.17226/20350.

5 3D Printing Vis-à-Vis Traditional Prototyping

5.1. MOVING AWAY FROM METAL

Polymers are commonly used because of their low density. Polymers have a very low strength compared to metals. Therefore, unless they have been reinforced with another substance, such as glass fibre or carbon fibre, polymers are not a particularly good choice for construction purpose.

Because they are lightweight and less expensive to create and change, polymers are widely used in daily life. If every door, table, chair, piece of cooking equipment, etc., were made of metal, it would be difficult to install everything where we wanted it. The weight of metal alone has significant disadvantages. Polymers are more affordable to produce than materials like copper or aluminium, for example. Steel is unquestionably less expensive than polymers, but you wouldn't want to use it for every single function. However, because of their exceptional strength, composites (polymer plus reinforcement) are increasingly sought-after in commercial applications.

Polymers have essentially replaced metals, wood, and other natural fibres in their primary usage due to their inherent benefits of reduced weight and greater water resistance [1] (excluding Nylon, which is used to make Kevlar).

There are many applications . . . let's say, for example, Kevlar.

Kevlar is essentially polyamide. The primary usage of polyamide is as a bulletproof jacket and other similar applications since all polyamides have extraordinarily high resistance to impacts due to the existence of aliphatic groups.

Speaking of replacing metals, the initial body armour and shields used in battle were all comprised of metal or leather. Metals were just too weighty and would slow down warriors, and also possessed the strength attributes to stop a bullet at such a thin thickness, therefore defences had to evolve as weapons did from swords and arrows to guns and rifles.

Now let's talk about the additional uses for polymers that replace metals. Starting with the automobile sector: Because they possess the required features, such as a decent impact resistance, colorability, good load bearing capacity, etc., metals were formerly thought to be the most ideal material for vehicle bodywork.

Today, however, automotive bodywork is made from composites or mixes made of PC (poly carbonate) and ABS (acrylonitryl butadiene styrene polymer), due to the higher average fuel consumption savings that their low weight would produce. It is possible to make increases in weather resistance, colorability, impact resistance, shock resistance, and attainable speed while keeping the engine capacity the same.

We could give many more instances, such as furniture or plastic water bottles. Polymers, either new or recycled, have essentially taken the place of other traditional materials since the beginning of the century.

DOI: 10.1201/9781003349341-5

5.1.1. MEDICAL AND LIFE SCIENCES

- **Sterility:** When it pertains to equipment, hygiene is crucial in the medical sector. Hospital patients and health care workers are most at risk from infection. Compared to metal, polymer and composites are simpler to clean and sterilise.
- **Radiolucency:** Radiant radiation such as X-rays can travel through radiolucent materials even while encountering some resistance. The surgeon can see clearly and clearly under fluoroscopy thanks to surgical equipment and components made of polymer materials. This enables safer, more accurate surgical results in the operating room. Metal tools obstruct the surgeon's vision.
- **Lightweight:** Orthopedic OEMs can comply with optimal weight requirements for surgical trays thanks to plastic and composite surgical components. The medical team must carry and use more metal devices, which increases weight and strain.
- **Reduced Stress-Shielding:** When metallic implants and bone don't really converge or function together, stress shielding takes place. However, in medical-grade polymeric materials like PEEK, the bone-like modulus "fuses" into a unified framework.

5.1.2. AEROSPACE AND DEFENSE

- **Lightweight:** Composite and polymer materials can be up to ten times lighter than conventional metals. The financial performance of an aircraft firm can be significantly impacted by part weight reduction. The airline may save up to $15,000 annually in fuel costs for each and every pound of reducing weight on a jet.
- **Corrosion-Resistant:** In settings with severe chemicals, plastic materials perform far better than metals. This lengthens the lifespan of the aircraft and prevents expensive repairs caused by corroding metallic components, which in turn reduces MRO downtime and improves the amount of time each aircraft is operating each year.
- **Insulating and Radar Absorbent:** In addition to being inherently thermally and electrically insulating, polymers also absorb radar.
- **Flame and Smoke Resistances:** The strict flame and smoke resistance standards necessary for aircraft applications are met by high-performance thermoplastic polymers.

5.2. CAUSE OF CHOOSING POLYMERS OVER METAL

5.2.1. MACHINED POLYMER AND COMPOSITE COMPONENTS ARE THE MOST COST-EFFECTIVE SOLUTION WHEN COMPARED TO METAL

Machined plastic components have a substantial benefit over metal components owing to their lighter weight in that they offer reduced lifetime shipping costs for equipment that is regularly moved or handled during the course of the product throughout its life cycle. Due to the lower frictional properties of polymer wearing parts as opposed to metal wearing parts, polymers provide substantial advantages over metals in bearing and wear applications. This enables lower energy motors for moving components. As a result of the reduced frictional properties, less wear is also achievable. Because to the decreased wear rates, maintenance is important less downtime. You'll earn more money now that your equipment can remain online for longer. Plastic products are not only significantly less costly than some of the raw metal pieces used to produce parts, but they are also more lightweight.

5.2.2. Plastics Are More Resistant to Chemicals Than Their Metal Counterparts

Metals are quickly harmed by numerous common substances when they lack extensive and costly surface coatings and other treatments. Corrosion of metal parts caused by moisture or simply by clashing metals is a serious problem. Examples of polymeric materials and composites that really are impervious to some of the most dangerous chemicals include PEEK, Kynar, Teflon, and polyethylene. Because of this, the chemical and manufacturing industries may now develop and employ precise fluid handling equipment that would normally dissolve if built of metallic materials. Some materials made of machinable polymers may withstand temperatures as high as 700°F (370°C). Whereas metal, plastic components don't need to be finished after the post-treatment process.

Materials that are electrically and thermally insulating include blends and polymers. Metal parts need to be processed and treated in an efficient and effective manner to obtain any type of insulating characteristics. These later phases increase the cost of metallic materials by not offering the same amount of insulation as polymer materials provide. Furthermore, when in touch with different metals, composite and plastic components don't require sheathing since they are naturally corrosion-resistant. Contrary to metals, polymers are coloured before being machined, eliminating the requirement for post-treatment finishing techniques like painting [2].

5.2.3. Other Advantages of Choosing Plastics Over Metallic Material

5.2.3.1. Design Flexibility

Plastics are created using a variety of resins. Despite the variations in each plastic resin, Fabriplastic polymers as a whole provide more flexibility than metals. Polymers are without a doubt a superior option for industries that demand a wider range of forms, textures, and geometries. Plastics have a number of crucial advantages over metals, including their ease of moulding while offering durability on par with metal.

When industrial products require complex geometries, aesthetically attractive designs, or lightweight qualities, plastic is frequently selected over metal. Modern techniques like plastic injection moulding allow the manufacture of mould shapes and parts with incredibly complex geometries with the same efficiency as metals.

5.2.3.2. Recyclability

Plastic may be recycled. And this is perhaps one of the key advantages of using plastic rather than metal. Used plastics are more affordable for company owners since they can be melted and recycled several times.

Metal components that have been broken must be thrown away and replaced with new ones, but plastics may be reused and used to create new products. This functionality also shortens the time it takes to locate new resources for use in productions.

5.2.3.3. Life Span

The exceptional resilience of plastics against threats from chemicals, physical forces, weather, and the climate is one of the key advantages of using plastics over metals. Metals normally have a long lifespan, however this lifespan is shortened by their vulnerability to chemical and environmental harm. Plastics, on the other hand, are supposed to be more resilient and have a longer life expectancy.

5.2.3.4. Build Time and Complexity

Generally speaking, building a metal three-dimensional object takes more time than one made of a polymer. Plastic 3D printers can produce objects in thicker quantities while using less energy. Due to the high degree of intricacy required by 3D printers that use metal materials, the printing process could take a while. Plastic also needs complexity to be the optimal material for 3D parts. It does not undergo the same amount of inspection that metal does.

5.2.3.5. Safety

Metal components manufacturing has certain risks. Numerous collisions and injuries routinely occur in industries that employ metals in any form. However, polymers are a healthier substance due to their greater softness and sharper ends [3].

5.2.3.6. Design Inaccuracies

Another possible difficulty with 3D printing appears to be the method or equipment used, since some printers have inferior tolerances, which might lead to final goods that differ from the original design. Although this may be fixed in post-processing, it must be considered that this will increase the production time and expense [4].

Polymer/metal nanocomposites have recently attracted the attention of several research and industrial organisations in an effort to create innovative products or replace currently utilised materials. Zare et al. described the properties and uses of polymer/metal nanocomposites for biomedical applications in a variety of fields, such as robust and stable materials, conducting devices, sensors, and healthcare goods [5]. The mechanical characteristics of syndiotactic polystyrene (PS) nanocomposites comprising 3 weight percent of various nanofillers, such as multi-walled carbon nanotubes (MWCNT), nano-diamonds, Cu nanofibres, and Ag nanoparticles, were also documented by Papageorgiou et al. It is demonstrated that adding metal nanoparticles to PS matrix can result in adequate Young's modulus, elongation, and impact strength [6] (Perna et al.). Integrating 3D printing of polymer matrix composites and metal additive layer manufacturing: surface metallization of 3D printed composite panels through cold spray deposition of aluminium particles

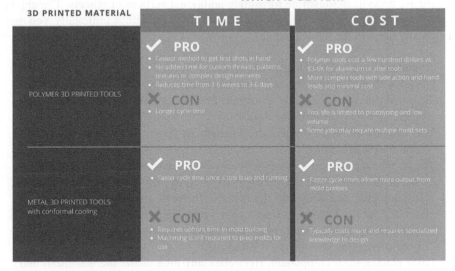

POLYMER VS METAL
3D PRINTED TOOLS
WHICH IS BETTER?

FIGURE 5.1 Polymer vs Metal Pros and Cons [8].

[7] Specifically, 3D-printed carbon fibre-reinforced plastic (CFRP) panels have been encased with aluminium particles to combining the light weight and superior mechanical properties of CFRP with maximum wear resistance and toughness of metals. This combines 3D printing of composites with cold spray deposition of metal particles. Fused filament fabrication (FFF) technology was utilised to create CFRP panels, which were subsequently coated with aluminium particles utilising a low-pressure cold spraying facility.

5.3. PRINTING COMPUTATIONALLY COMPLEX OBJECTS

Even when a 3D model is describing a physical print that easily fits into a printer's build volume, objects can occasionally have extremely complicated 3D models. As we shall examine in Chapter 13's treatment of 3D-printed representations of complicated molecules, this issue, for instance, occurs in certain scientific visualisations. It is alluring to want to preserve every little detail, but a consumer-level printer will likely not do so in the case of a very complicated model. Cutting the model into manageable-sized sections, then slicing and printing each one independently helps reduce the STL file sizes of the parts if the model is both physically large and very complex. With MeshLab, a free open-source tool, you may browse, combine, transform, or fix point clouds in addition to a variety of file kinds, including STL, PLY, OFF, OBJ, and 3DS. Visit the Sourceforge project page to obtain MeshLab without charge.

Step 1: Open STL file

To open a supported mesh file go to **File > Import Mesh** and browse for your model.

Step 2: Edit STL file

There are no tools in MeshLab for adding additional vertices or objects. The ability to mix two models from a 3D scan is a fantastic tool, though. Additionally, you may patch any holes in the model and delete portions of the mesh.

To combine two meshes load both models.

Click on **Show Layers.** Select the models in the scene.

To transform, rotate or scale an object, first select it in the **Layer** menu and then click on **Manipulator Tools**.

You can now press **T** to select the **Transform** option, **R** to rotate the model and **S** to start scaling.

The coordinate system we are functioning in depends on the way you are looking at the model. You may easily resize or move the model in one direction by dragging and dropping the arrows. To rotate the item, turn the circle around it. Hit Escape to rotate your view, and when you have the ideal perspective on an object, press Escape once more to begin changing. To confirm the positioning, press **Enter**.

After placing each component, right-click any area of the merged mesh and select Smooth Visible Layers from the menu that appears. Click **Apply** after selecting the first three checkboxes.

5.3.1. HOW TO REPAIR YOUR OBJECT OR SEARCH FOR HOLES

*   Simply click Fill Hole. In order to use this option, your structure must be complex.
*   You will be shown every hole in the model in a window that will appear. Selecting the holes you want to fill is now possible. When chosen, they will glow green. To complete, click Fill and then Accept.

Step 3: STL Repair

*   To check if your STL file is watertight, select **Filters > Quality Measures and Computations > Compute Geometric Measures.** You should see a volume for your file or an error report in the dialog on the right side.
*   In case your STL file is not watertight, select **Filters > Cleaning and Repairing > Merge Close Vertices**, click **Apply.**
*   Select **Filters > Cleaning and Repairing > Remove Duplicate Faces**, click **Apply**.
*   Select **Filters > Cleaning and Repairing > Remove Duplicated Vertices.**

Step 4: Export as STL file

To export the model, go to **File > Export Mesh.**

Compressing an STL is a great approach to minimise the amount of storage space it requires, as it is with most files. Reducing your file will, however, only slightly

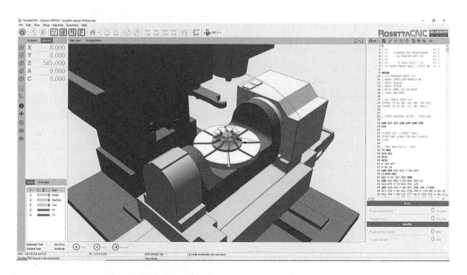

FIGURE 5.2 Slicer Repairing Interface [10].

improve performance when interacting with 3D files, such as STL (as opposed to 2D ones).

An STL file may be reduced by about 73% through compression. However, if you have a lot of details to retain, this proportion may occasionally be substantially lower. Therefore, compression is the way to go if you simply need to cut a few MB while not wanting to compromise the quality of your model.

To compress an STL file, you can use programs or web apps like *GZIP* or *7-Zip*. [9]

Tip: An STL file more than a few tens of gigabytes is likely too large for most printers to replicate as well as for most slicing methods to handle. Practically, you can only slice files up to about 50MB, or even less if you're using an old PC. Take into account the two methods covered in this section if you have a larger file.

5.4. PRINTING PHYSICALLY BIG OBJECTS

Because it is far more difficult to maintain temperature and accuracy of the build platform and the rest of the machine's construction, consumer-level 3D printer build areas are often just a small fraction of their professional counterparts. Because of this, the majority of consumer printer construct areas are less than 200 mm in any dimension, and many are much less. This implies that if you wish to make something larger than that, you will need to use software to split the item into parts, print individual parts, and then somehow assemble the whole thing from those parts.

You can chop an entity into pieces if you really want to print it but it won't fit in the construction volume. For instance, if the item is long and thin, you may cut it into multiple pieces, arrange them on the build plate, and print them all at once. If not, you will need to print the item in more than one run. In either scenario, you will have to glue or put the component together later.

5.4.1. OBJECTS THAT ARE TOO LONG FOR THE BUILD PLATFORM

Check to determine whether an item will fit diagonally on the substrate if it is long and slender and will not fit inside the longest construction dimension of your printer. There could be some trade-offs there as you'll also want to preserve the orientation that will reduce support. It could be simpler to cut the thing in half and afterwards join two halves that would both rest flat instead of picking off a significant amount of support if you plan to print diagonally in all three dimensions. If you do need to print a long, thin item (like a chopstick), you'll probably be able to put the parts close to one another on the platform and have them print at the same time. You might cut it into many pieces, arrange them on the construction floor in one or more clusters, and afterwards glue them together when you were attempting to print a tall, slender structure.

5.4.2. OBJECTS THAT ARE TOO BIG IN MORE THAN ONE DIMENSION

The item you are attempting to print must be divided along two axes, or at least into four parts with a lateral slice and a horizontal slice, if it is too large in much more than one dimension. Determine where your printer has the greatest precision and plan cuts so that any components that need a lot of precision—like those with important joints—will fall there. The middle of the print bed often has the highest resolution.

5.4.3. GLUING THE PIECES TOGETHER

Use adhesive made for plastics once you have printed the components. Adhesives made of cyanoacrylate, or "superglue," function rather well with PLA and ABS. With any household adhesive, nylon is challenging to bond.

5.5. POST PROCESSING

Now, finalizing the AM technique and matching for the intended application can be considered as a complement, if not a prerequisite, to the focus on post-processing and finishing. The steps below are included in AM, as was previously mentioned: Designing using CAD, exporting as an STL file, cutting, printing, and post-processing. The post-processing/finishing techniques used with 3D printed polymers will be thoroughly addressed and explained in this section. Support removal, powder removal, or resin removal are significant post-processing procedures in additive manufacturing that often take a lot of time. These operations fall within the principal post-processing category [11].

5.5.1. MATERIAL SUPPORT REMOVAL

This is carried out by the removal of complex support elements and is seen to be the most fundamental kind of post-processing. Simple needle-nose pliers or flush cutters are all that are needed for this procedure. Simply cutting, shaving, or deburring extra material does this. Additionally, even though it might be labor- and time-intensive, it doesn't change the part's overall shape. On occasion, though, this procedure leaves traces on the component's surface. Some 3D printers combine AM and SM to avoid this procedure (machining or polishing combined). A water-soluble support, such as polyvinyl alcohol (PVA), is extruded by the secondary extrusion nozzle in some 3D printers that feature twin extruders. PVA may be easily eliminated by dissolving it in a water solvent. Fused deposition modelling prints as well as components made with other printing processes can have their material support removed.

5.5.2. Sanding or Surface Polishing

After the supports have been taken out, sanding is done. In this procedure, any noticeable flaws in a 3D item are eliminated by smoothening it with a sander or sandpaper [11]. In order to eliminate bumps and scratches from the print, it is advised to apply increasing grades of sandpaper (i.e., 100, 240, 400, 600, 1500, and 2000). Although this technique could result in a pleasing texture, it is challenging to apply to complex surfaces or lower glass transition (Tg) polymeric materials.

5.5.3. Gap Filling

This makes use of readily available, affordable materials to at the very least preserve and enhance an FDM print surface. Gaps may result from incompletely printed layers due to toolpath restrictions and other factors. Simply said, gap filling involves adding epoxy or other fillers to spaces and gaps. For bigger gaps that need additional sanding to eliminate more material, a filler may be employed.

5.5.4. Coating

This is the application of any paint or resin mixture that would enhance the print's appearance. Manual painting can be done using a brush or air spray. This technique is relatively easy to use, much like the other post-processing techniques mentioned above. The findings of the investigation demonstrated appropriate mechanical characteristics for specific applications (particularly non-load bearing), which depicts 3D-printed samples that have been coated to evaluate its influence on strength and stiffness of the material. The coating polymers polyurethane elastomer and liquid silicone were both employed. The reference has the specifics. These two materials are frequently used to weatherproof outdoor spaces.

5.5.5. Polymer Coating

Direct polymeric coating, in contrast, is discovered to boost the adherence of 3D printing materials to textile fibres by first coating them with a soluble polymer layer. Coating is often done using spray paints and other formulations for aesthetic purposes (dissolved or melt polymer). One of its benefits is that adhesion may be greatly improved while maintaining the fabric's haptic qualities and bending stiffness. Plastisols are polymer-solvent solutions that must be coated before the solvent evaporates. Epoxy resins are thermoset coatings made of polymers that must be cured [11, 12].

5.5.6.   UV Post-Curing

The thorough curing of the residual resin during this post-processing increases the robustness of 3D printing. Additionally, post-curing at higher temperatures could speed up the curing process and produce greater mechanical characteristics. However, it takes a lot of time and calls for certain lab settings. Additionally, it's important to appropriately dispose of solutions, especially those used to dissolve resins [3, 11]. Be aware that some businesses create post-curing equipment specifically for their 3D printers. Garcia et al. investigated how UV post-curing affected the mechanical characteristics of SLA-printed items using the Form 2 SLA 3D printer and the industry-standard clear resin (both from Formlabs).

5.5.7.   Thermal Post-Curing

According to Uzcategui et al., the most popular post-processing techniques used to enhance conversion and the mechanical characteristics following SLA printing are UV curing and thermal curing. They did, however, emphasise that the inner parts of 3D-printed parts do not benefit from UV curing, which accounts for their lower qualities when compared to materials that have undergone complete conversion. The "candy-shell" effect brought on by the presence of an absorber is the cause of this. As a result, just a thin layer becomes hardened (i.e., the outer surface only). To induce high conversion over the whole material, the scientists suggested using a thermal activator during a thermally post-cure [11].

5.5.8.   Other Finishing Methods

For 3D-printed items, there are also other, more advanced techniques: (1) Physical vapour deposition (PVD) involves ionising the atoms of the coating material to cover the surface of the 3D-printed item with a metal or ceramic substance. The coating can be transferred using this method without the need for a medium, however it is often carried out in a vacuum. These coatings alter the surface's properties, resulting

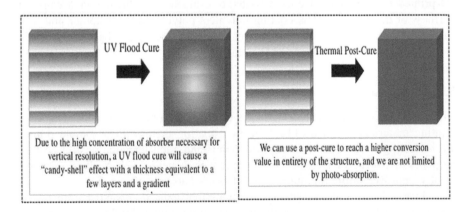

FIGURE 5.3   UV and Thermal Post-Curing [11].

in a component that is more resistant to heat, chemicals, friction, wear, and other factors.

(2) Common 3D printing polymers may have their surface qualities functionalized via a process called chemical vapour deposition (CVD), which makes it possible to alter and design the bulk properties of 3D-printed objects—like strength—independently from the surface properties. In one investigation, the pieces were coated using CVD while the filament temperature and substrate temperature were both altered (hydrophilic and hydrophobic polymers were used as a coating). A schematic of the initiated chemical vapour deposition (iCVD) procedure used to cover PLA and ABS 3D-printed objects and substrates. After being set up on a silicon wafer portion and a temperature-controlled stage, the white 3D-printed substrate was installed. The substrates were coated with both the hydrophilic poly(2-hydroxyethyl methacrylate)-co-(ethylene glycol diacrylate)) (P(HEMA-co-EGDA)) and the hydrophobic poly(1H,1H,2H,2H-perfluorodecyl acrylate) (PPFDA). The lattice's pores stayed dry and filled with air because the uncoated PLA is hydrophobic, which decreased the density of the lattice and prevented it from sinking. Similar to that, the PPFDA-coated item floated because its increased hydrophobicity prevented its pores from being wet, causing them to be filled with air. In contrast, the item coated with P(HEMA-co-EGDA) sunk in the water because its hydrophilicity allowed water to enter its pores. This sample was submerged in water for three days, dried, and then reinserted into the water, where it once more sunk. This test showed that its hydrophilicity may be preserved. These findings demonstrated that these 3D-printed functionalized components might be used to tissue scaffolds and microfluidics [12]. A hydrophilic polymer was applied to 3D-printed pieces first in another experiment, followed by a hydrophobic polymer. ABS material was used to print the components and substrates. The iCVD procedure was used to coat several components, including a bolt, nut, and comb demonstrates how hydrophobic the uncoated ABS component surfaces were. The hydrophilic P(HEMA-co-EGDA) coating was applied initially, and these components were easily moistened. Following that, the components were covered in PPFDA, and it was then possible to see that the surfaces had restored their hydrophobicity These tests demonstrate how the CVD method may be used to tailor the surfaces of 3D-printed objects.

(3) The technique of electroplating involves using the electro-deposition method to apply a thin layer or coating to the surface of another object for ornamental or practical purposes. In reality, numerous metallic alloys may be electroplated onto 3D-printed items produced using SLA, FDM, Polyjet, and other methods. The surface would have improved qualities in this situation, including enhanced electrical characteristics, higher strength, improved heat deflection, enhanced chemical resistance, aesthetic value, and a smooth finish.

(4) The elimination of metallic ions from a liquid electrolyte results in the homogenous coating of broad metallic layers on the base of non-conductive materials (substrates), a process known as electroless deposition or plating, also known as autocatalytic deposition of metals. This technique was utilised in certain articles to coat nickel platinum, copper, and palladium metallic materials for 3D-printed components. The researchers also noted that the components with functionalized surfaces might be used to various chemical and mechanical engineering systems, as well

as to MEMS, microrobots, and metamaterials in the realm of electronics. Through electroless deposition, Jones et al. plated palladium onto 3D-printed photopolymers. The authors created a cylinder with cubic lattices and printed cubes with logpile lattices. The tests made use of the PR48/PR57 and an Autodesk Ember 3D printer. The cubes had 200 m pores and measured 5 mm along its edges. The cubic lattice was made up of 150 m-wide holes, and the cylinders had dimensions of 200 mm in length and 2 mm in diameter. One hour was spent plating with gentle stirring. The substrate was air dried after being cleaned with deionized water. Other thermoplastic materials produced with FDM showed similar adhesiveness, film brilliance, and plating rate. Comparatively speaking to FDM-printed components, particularly polycarbonate, internal plating of photopolymers might be easily accomplished. These findings demonstrate that 3D printing and a straightforward electroless deposition technique may be used to make 3D metal–polymer composite objects quickly and efficiently (essentially creating a functional metal surface on high-resolution polymer substrates). This might lead to the creation of useful 3D parts that make use of mechanical, electrical, thermal, and catalytic capabilities.

(5) Protective post-processing against electrostatic discharge (ESD). Component materials for many electric/electronic devices must be electrostatic discharge (ESD) -safe in order to avoid injury from the accumulation of electric potential (static discharge). Typically, three post-processing techniques are used: painting or coating; covering with conductive tape; and wrapping with films that are aluminum- or carbon-coated. There is no longer a requirement for post-processing thanks to the development of ESD-safe 3D printing materials by several businesses.

5.6. TROUBLESHOOTING

Troubleshooting methods often try to focus on an issue so that it may be explored. Finding the problem and testing quick remedies like restarting the computer, turning the gadget on and off, and ensuring the power cable is connected are the main goals.

Troubleshooters start by looking for usual, well-known reasons. When a laptop won't load, for instance, the power cord should clearly be checked first. Once common problems have indeed been ruled out, troubleshooters must look through a list of parts to identify the portion where the breakdown is occurring [13].

The basic aims of troubleshooting are to identify the cause of something not working as intended and provide a repair for the issue.

This tutorial is a great place to start if you want to improve the quality of the 3D printed objects you produce. The software choices you may utilise to solve the most frequent 3D printing issues have been thoroughly outlined.

5.6.1. Not Extruding at Start of Print

This issue is pretty common among novice 3D printer users, but thankfully, it is also relatively easy to resolve. There are actually four possible causes if the extruder doesn't really start to flow plastic at the beginning of your printing. We'll go through each one and explain how to fix the problem shortly.

5.6.1.1. Filament Was Not Loaded Into Extruder Before Printing

Most extruders have a propensity to leak plastic when left standing in hot environments. The heated plastic from the inside of the nozzle tends to leak out of the tip, leaving a hole where it formerly was filled with plastic. When your extruder is being warmed at the beginning of a print or when it is progressively cooling at the conclusion of the print, you may see idle oozing. Even if your extruder has wasted some plastic due to oozing, it should take a little while the next time you try to extrude until plastic starts to flow through the nozzle once more. You can see the same slow extrusion if you try to continue a print after your nozzle has started to ooze. To solve this issue, prime your extruder right before beginning a print so that the nozzle is ready to extrude and filled with plastic. In Cura, this is usually done by including a garment called a skirt. The skirt will wrap your component, feeding plastic into the extruder in the process. If you require additional priming, increase the number of skirt outlines in Cura's Expert tab. To release plastic before printing, users may also utilise the control knob on a 3D printer.

5.6.1.2. The Distance Between Nozzle and Bed Is Too Close

If the nozzle somehow gets too near to the platform, there won't be enough room for plastic to escape the extruder. The top hole of the nozzle is virtually closed off, preventing any plastic from leaving it. It is simple to identify the issue if the print doesn't extrude plastic for the first layer or two but begins to do so regularly around the third or fourth layer. The four screws in the four corners of the hot bed can be turned clockwise to fix this problem, causing the hot bed to drop. Turn a little bit after extending the spacing once until there is enough place for plastic extrusion.

5.6.1.3. The Filament Stripped Against the Drive Gear

3D printers push the filament back and forth using a small gear. Because the teeth on this gear bite into the filament, they can accurately control where the filament is located. However, if you see a lot of plastic shavings or it looks like some of your filament is missing, the driving gear may have removed too much plastic. The driving gear will therefore have nothing to cling onto when it tries to drive the filament back and forth between when this happens. This issue requires the following solutions:

- **Adjust extruder tension:** Extruders frequently include one or two screws, which you can adjust to change how firmly the filament is gripped by the extruder's gear. You can find an example of what we're talking about in Figure 15. If it's too tight, loosen the filament's grip on the feeder of your printer (gear component).
- **Replace the PTFE tube:** If your Bowden tube or hot end coating (which is frequently a PTFE tube) are restricting filament flow, you might have to replace them. Make absolutely sure it may not be twisted or bent in any manner.
- **Reduce retraction:** If the Bowden tube or hot end coatings (which is frequently a PTFE tube) are restricting filament flow, you might have to replace them. Make absolutely sure it may not be twisted or bent in any manner.

- **Change speed**: Striping might result from printing too gradually or too rapidly. To combat the former, several manufacturers suggest accelerating printing to avoid heat creep and hot end clogs. Try reducing print speed in the latter case to avoid overtaxing the extruder.
- **Adjust hot end temperature:** It's possible that you believe your filaments aren't extruding quickly enough. In that case, warm the hot end to the standard printing temperature and carefully feed some filament (but not too fast). The pace at which the filament extrudes should closely match the pressure. If you notice that the filament isn't really melting quickly enough, raise the nozzle's temperature.

5.6.1.4. The Extruder Is Clogged

The extruder is likely clogged if any of the previous recommended fixes are unsuccessful in solving the issue. This can happen when foreign items become lodged in the nozzle, when molten plastic sits within the extruder for a long time, or whenever the thermal cooling of the extruder is inadequate and the filament begins to soften outside the proper melt zone. Using relative diameter drill (or relative diameter copper line of electric wire), we were able to clear clogged extruders. The extruder has to be heated to 200 °C before performing this. Focus is unaffected by the hot extruder. Please contact the maker of your printer if you require repairing the extruder.

5.6.1.5. Print Not Sticking to the Bed

In order for the successive layers of the product to be built on top of the initial layer, it must be completely connected to the build plate of the printer. If the initial layer doesn't stick to the build plate, later problems will occur. We'll look at a few typical reasons of first-layer adhesion problems in the next section, along with solutions for each.

5.6.1.6. Build Platform Is Not Level

A 3D printer's bed may be moved with four screws. If you're having difficulty getting your initial layer to stick to the bed, make absolutely sure the printer's bed is calibrated and flat. If the bed is not leveled, one side may be very near to the nozzle while the other may be rather far away. For the best first layer, a steady print bed is required.

5.6.1.7. Nozzle Starts Too Far Away From the Bed

Once the bed has truly been correctly levelled, you should still verify that the nozzle is starting at the correct height in respect to the building platform. Your extruder should be positioned so that it is neither too far away from nor too near the build plate. For optimal adhesion, your filament should be just a tiny bit sticky to the build plate. The four screws found in the heated bed's corners may be turned counterclockwise to elevate the heated bed. When there is sufficient space for filament extrusion, turn a bit to close the gap (the distance is about the thickness of a piece of A4 paper). Use extreme caution whenever you change this setting. Since the part's layers are generally only 0.2mm thick, even a small alteration can have a significant effect.

5.6.1.8.   The Build Platform (Tape, Glues, and Materials)

Various polymers tend to adhere to different materials more readily. If you wish to print directly onto these surfaces, it is often a good idea to make sure your platen is free of dust, oil, or lubricants before starting the print. Cleaning your print bed with mild water or isopropyl rubbing alcohol may be quite helpful. Tape of all varieties sticks firmly to common 3D printing materials. Tape strips may be added to the build platform interface and readily removed or replaced if you wish to print with a different material. For instance, PLA usually sticks to blue painter's tape well whereas ABS usually sticks to Kapton tape better (commonly known as Polyimide film). Their building bases can have temporary glue or spray applied to the tops of them. Hairspray, glue sticks, and other sticky products frequently function pretty well if all else fails. Try a few different things to see what suits you best.

5.6.1.9.   Brims and Rafts

There may not always be sufficient surface area for an item to attach to the construction platform when printing extremely tiny pieces. Cura has a number of options that may be used to increase this surface area, providing the print bed a larger area to cling to. One of these options is called "Brim." The brim increases the number of rings on the exterior of your component, much as how a hat brim widens the hat's circumference. This feature may be activated by choosing the "Support" page and turning on the "Platform adhesion type" option.

5.6.1.10.   Not Extruding Enough Plastic

How much material the 3D printer should discharge is managed by a few settings in Cura and Creality Slicer. However, because the 3D printer provides no information on how much material really emerges from the nozzle, it's possible that less plastic is emerging from the nozzle than the programme predicts (otherwise known as under-extrusion). In this scenario, gaps between succeeding extrusions of each layer may be visible. The most accurate test you can do to see if the printer is extruding sufficient plastic is printing a simple 20mm tall cube with at least three perimeter outlines. Make sure the cube's three edges are securely bonded at the top. If any of the three perimeters have gaps in them, users are under-extruding. If none of the gaps exist and all three perimeters touch, you are likely confronted with a different issue. You may be under-extruding for a variety of causes, some of which we have included here.

5.7.   PRINTING DEFECTS AND REMEDIES

5.7.1.   Incorrect Filament Diameter

The programme must first be verified to be aware of the filament diameter you are using. Where can you locate this setting? Under "Basic," on the "Filament" tab. Check to see whether this value matches the filament you purchased. You may even wish to use a set of callipers to measure your filament in order to confirm that the diameter indicated in the application is accurate. The most common filament diameter ranges from 1.75mm to 2.85mm. The diameter of the filament used in 3D printers is 1.75mm.

5.7.2. INCREASE THE EXTRUSION SPEED

If the filament diameter is suitable but you are still having issues with under-extrusion, you have to modify your extrusion frequencies. This option in the 3D printer control panel allows you to quickly modify the amount of plastic extruded (otherwise known as the flow rate). The "Configuration" knob can be turned to access this parameter. This is how exact setup is done: Move the knob to enter the menu, choose "Configuration," then "Confirm," and then click to the "motion," choose the "Esteps/mm" tab, and then, lastly, turn the knob to change the extrusion speed. The primary key may be changed from 95mm/s to 97mm/s. It suggests that the pace of plastic extrusion will rise by 2 mm/s. Increase the extrusion rate as well before printing the test cube again to see whether you still have gaps in your perimeters. Be note that this increment was only implemented once to avoid plastic being extruded too frequently.

5.7.3. HOLES AND GAPS IN THE TOP LAYERS

In order to save plastic, the majority of 3D printed things have solid exteriors concealing porous, partially empty interiors. For example, a 30% infill percentage would imply that only 30% of the interior would've been totally composed of solid plastic and the other 70% would've been made up of air. Even if the interior of the part may be somewhat hollow, we want the outside to remain solid. To do this, you may specify how many solid layers you want on the top and bottom of any component in 3D mode. For instance, the technique would generate five solid layers totally at the top and five somewhat hollow layers throughout the rest of the print if you were printing a simple cube with five solid layers at the top and five solid layers just at the bottom. This process may create components that are highly durable while reducing the amount of time and plastic used. However, depending on the parameters you are using, the top solid layers of your print might not be totally solid. You could see cavities or holes in between the extrusions that make up these rigid layers. If you've already encountered this issue, you may alter a couple of these simple settings to fix it.

5.7.4. NOT ENOUGH TOP SOLID LAYERS

The initial parameter that has to be altered is the number of highest solid layers. Anytime you try to build a 100% solid layer on top of your partially hollow infill, the solid layer must bridge the empty air pockets of your infill. When this happens, the extrusions for the solid have a tendency to droop or sink into the air pocket. You frequently need to print multiple solid layers just at top of the print in order to have a great flat, completely solid surface. As a general guideline, the solid component at the top of your print should be at least 0.6mm thick. If you were using a 0.2mm layer height, you would consequently need at least three top solid layers. You could need six solid layers just at top of your print to achieve the same effect when printing at a lower layer height, such as 0.1mm. If you are 18 and you see gaps between the extrusions on the upper surface, you should start by increasing the number of top solid layers. To test if the problem is fixed, try printing with 1mm solid layers rather than only 0.6mm solid layers. Remember that additional solid layers will occur over the whole dimension of your component; they do not cause the part's outside to become larger.

5.7.5. INFILL PERCENTAGE IS TOO LOW

The layers that will be applied on top of the internal infill of the portion will rest on it. The top solid layers of the component must be printed on top of this foundation. If your infill percentage is really low, there will be large air spaces in your infill. For instance, if you only used 10% infill, 90% of, say, the inside of each item would be empty. As a result, the solid layers would have to print across some rather sizable air gaps. If you've tried increasing the number of top solid layers but are still observing gaps in the print's peak, you might wish to try increasing your infill percentage to see if they go away.

5.7.6. STRINGING OR OOZING

Small threads of plastic are left behind on a 3D printed item and are referred to as stringing, also known as oozing, whiskers, or "hairy" prints. This often happens when the extruder moves to a new spot and plastic leaks from the nozzle. Fortunately, Cura has a number of options that can resolve this problem. Retraction is the most typical setting that is utilised to prevent excessive stringing. When retraction is enabled, the filament will be dragged backwards into the nozzle to prevent oozing once the extruder has finished printing one area of your creation. The filament will be forced back into the nozzle when it is time to start printing again, causing plastic to start ejecting from the tip once more. Make sure retraction is enabled by selecting "Basic" and then "Quality." Make sure your extruder's retraction option is turned on. In the sections that follow, we'll go over crucial retraction parameters as well as a number of additional options for preventing stringing, such as extruder heating settings.

5.7.6.1. Retraction Distance

The retractable distance is the most crucial retraction setting. How much plastic is drawn out of the nozzle depends on this. To set the retraction distance in Cura, select "Advanced" and then the Retraction tab. Generally, the nozzle is much less likely to leak while moving, the more material is retracted from it. The majority of direct-drive extruders only need a 3.5 mm retraction distance. Try increasing the retraction distance by 0.5mm each time you run a test to see if the efficiency increases if you experience stringing with your prints.

5.7.6.2. Retraction Speed

The speed of retraction controls how quickly the filament is drawn back from the nozzle. If you retract too slowly, the plastic may begin to slowly seep from the nozzle and begin to move to its new location before the extruder is finished moving. If you retract too fast, the drive gear may keep chipping away some of the filament or the filament may split from the hot plastic within the nozzle. Retraction normally functions best in the sweet region between 30 and 50 mm/s. Slicing software has already provided a large number of pre-configured profiles that can serve as a starting point for determining what retraction speed is ideal. However, the ideal speed depends on the material you are using, so you may want to experiment to determine whether changing the speed reduces the amount of stringing you observe.

5.7.7. Mechanical or Electrical Issues

Even after slowing down the print speed, if the layer misalignment persists, there are probably mechanical or electrical problems with the printer. For instance, the majority of 3D printers employ belts to enable the motors to regulate the nozzle position. The belts are normally constructed of rubber and strengthened with steel wire to offer more durability. These belts might stretch over time, which could affect how tight of a belt tension is needed to hold the nozzle in place. The belt may slide onto the driving pulley if the tension gets too loose, which would result in the pulley turning but the belt remaining stationary. Initially installing the belt too tightly might also result in problems. A belt that is too tight might cause excessive bearing friction, which can stop the motors from turning. A belt that is just tight enough to avoid slippage but not so tight that it prevents rotation is ideal for assembly. In order to avoid problems with misaligned layers, make sure that all of your belts have the right tension and don't look to be either too tight or too loose. Please ask the printer's maker for information on how to change the belt tension if you suspect there could be a problem. Additionally, a lot of 3D printers include a number of belts that are powered by pulleys that are connected to stepper motor shafts by a little setscrew (otherwise known as a grub screw). These setscrews secure the pulley to the motor shaft, allowing the two components to rotate simultaneously. The pulley will no more revolve with the drive shaft if the setscrew becomes loose. This indicates that although the engine may be turning, the pulley and belts are not. This prevents the nozzle from reaching the target spot, which may affect the alignment of all next print layers. Therefore, you should make sure that each of the motor fasteners are correctly tightened if layer misaligned is a persistent issue.

The motors may also become displaced due to a number of other typical electrical problems. For instance, if the motors are not receiving enough electrical current, they won't have enough power to spin. Additionally, it is conceivable for the motor driver electronics to overheat, which would halt the motors from rotating until they cool down. Even while this is not a comprehensive list, it offers some suggestions for typical electrical and mechanical issues that you might wish to investigate if layer shifting is a recurring issue.

5.7.8. Clogged Extruder

Over the course of its lifespan, your 3D printer must melt and extrude several kilos of plastic. All of this plastic must escape the extruder through a small hole that is little larger than a single grain of sand, which complicates matters further. Eventually, something may go wrong with this procedure and the extruder might no longer be able to force plastic through the nozzle. Usually, anything inside the nozzle is the cause of these jams or blockages because it prevents the plastic from freely extruding.

While this could seem difficult the first time it occurs, we'll go through a few simple troubleshooting techniques that can be applied to unclog a blocked nozzle.

5.7.8.1.   Manually Push the Filament to Extruder

Manually inserting the filament into the extruder could be one of the first actions you need to attempt. Heat your extruder to the proper temperature for your plastic by opening the 3D Dashboard. Next, eject a little amount of plastic, say 10mm, using the Control knob. Lightly push the filament into the extruder with your hands as the extruder motor is rotating. This additional force will frequently be sufficient to move the filament past the problematic location.

5.7.8.2.   Reload the Filament

The following step is to unload the filament if it is still not moving. Make sure the extruder is heated to the proper temperature, then swiftly insert the filament into the extruder and pull it out. As previously, you might need to use a little more effort if the filament isn't moving. After the filament has been removed, cut away the melted or damaged piece of the filament using a pair of sharp scissors. Reload the filament after that to test if the fresh, undamaged segment of filament may be used for extrusion.

Layer Separation and Splitting
Layers are separating and splitting apart while printing

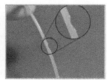

Grinding Filament
Plastic is being ground away until the filament no longer moves, otherwise known as "stripped" filament

Clogged Extruder
Extruder is clogged or jammed and will no longer extrude plastic from the nozzle tip

Stops Extruding Mid Print
Printer stops extruding plastic randomly in the middle of a print

Weak Infill
Very thin, stringy infill that creates a weak interior and does not bond together well

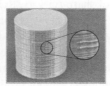

Blobs and Zits
Small blobs on the surface of print, otherwise known as zits

Gaps Between Infill and Outline
Gaps between the outline of the part and the outer solid infill layers

Curling or Rough Corners
Corners of the print tend to curl and deform after they are printed

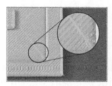

Scars on Top Surface
The nozzle drags across the top of the print and creates a scar on the surface

Gaps in Floor Corners
Gaps in the corners of the print, where the top layer does not join to the outline of the next layer

Lines on the Side of Print
Side walls are not smooth, lines are visible on the side of the print

Vibrations and Ringing
Vibrations that cause oscillations on the surface of the print, otherwise known as "ringing"

FIGURE 5.4   Printing Defects [14].

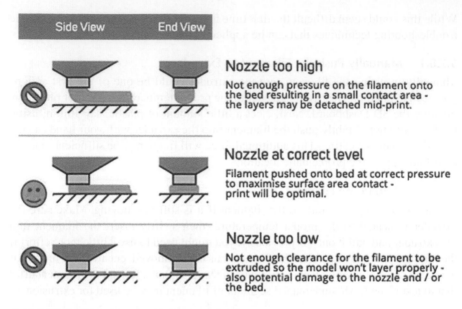

FIGURE 5.5   Nozzle and Bed Level of 3D Printer [15].

5.8. SMALL FEATURES PRINTING

5.8.1. Very Small Features Not Being Printed

The following step is to unload the filament if it is still not moving. Make sure the extruder is heated to the proper temperature, then swiftly insert the filament into the extruder and pull it out. As previously, you might need to use a little more effort if the filament isn't moving. After the filament has been removed, cut away the melted or damaged piece of the filament using a pair of sharp scissors. Reload the filament after that to test if the fresh, undamaged segment of filament may be used for extrusion.

Always make sure that the extrusion width is larger than or equal to the nozzle diameter. As a result, you could notice that Cura excludes these minor elements from the preview when you select "Layers." This is the software's method of alerting you that the nozzle on your 3D printer at the moment cannot produce these really little details. This can be a persistent problem for you if you routinely print tiny pieces. You have a variety of alternatives for effectively printing these tiny pieces. Shortly, we will give an example of each.

5.8.2. Redesign the Parts With Thicker Features

Redesigning the component to only contain features larger than your nozzle diameter is the first and most obvious solution. The size of the minor features is often altered by modifying the 3D model in the original CAD programme. After adding thickness to the minor details, you may load the model again into Cura to make sure your printer can print the 3D shape you made. If the features can be seen in the Layers, the printer will be able to reproduce the updated features.

5.8.3. Install a Nozzle With a Smaller Tip Size

In many instances, it is impossible to alter the original 3D model. It can be a component that you acquired from the internet or one that was built by someone else. In this situation, you might want to think about getting your 3D printer a second nozzle so it can produce tiny features. The nozzle tip can often be removed from printers, which makes these aftermarket changes quite simple. For instance, to give two alternatives, many customers buy a 0.3mm and a 0.5mm nozzle. For further information on how to install a lower nozzle tip size, speak with the maker of your printer.

5.8.4. Force the Software to Print Small Features

You have another choice if you are unable to modify the original 3D model and your 3D printer cannot accommodate a smaller nozzle. The programme may be made to print these inconspicuous details. But it's likely that this will have an impact on print quality. You may manually choose the extrusion width that the software should be using for your printer by selecting "Advanced" and then "Nozzle size." To force the programme to print elements as small as 0.3mm in size, if you had a 0.4mm nozzle, you might choose a manual extrusion width of 0.3mm. However, as we already indicated, most nozzles are unable to correctly produce an extrusion that is smaller than the tip size. As a result, keep a close eye on your printer to make sure the quality is suitable for these delicate details.

5.8.5. Inconsistent Extrusion

To create precise components, the printer has to be able to extrude plastic at a somewhat constant rate. If this extrusion varies in different parts of your print, it will have an impact on the finished print quality. Pay particular attention to your printer when printing to look for erratic extrusion.

For instance, if the printer is producing a 20mm-long straight line, but you observe that the extrusion is rough or varies in size, you are probably dealing with this problem. We have listed the most frequent reasons for uneven extrusion along with solutions for each.

5.8.6. Filament Is Getting Stuck or Tangled

The spool of plastic that is going into the printer's head should be checked first. Make sure that perhaps the plastic can be unwound from the spool with ease and that the spool can rotate freely. The thread will not be extruded through the nozzle uniformly if it becomes twisted or the spool encounters too much resistance to rotate smoothly.

5.8.7. Clogged Extruder

The next item to examine is the nozzle itself to see if filament is really not snarled and can be readily pushed into the extruder. If correct extrusion is being hindered, it's conceivable that there is some little debris or foreign plastic within the nozzle. Using the control panel of a 3D printer to manually discharge some plastic from the nozzle is a simple method to verify this. Make sure the plastic is extruding regularly and evenly by keeping an eye on it. If issues arise, you might need to clean the nozzle. For information on how to thoroughly clean the interior of the nozzle, please contact the manufacturer [13, 14].

REFERENCES

1. Mather, R. R. (2011). High performance fibres. In *Chemistry of textile fibres.* Royal Society of Chemistry.
2. MacDonald, J. (2021, January 7). Advantages of polymer components over metallic materials for machined parts. *AIP Precision Machining.* https://aipprecision.com/the-advantage-of-plastics-over-metals-for-custom-parts/
3. www.unipipes.com/blog/plastics-over-metals
4. www.twi-global.com/technical-knowledge/faqs/what-is-3d-printing/pros-and-cons
5. Zare, Y., & Shabani, I. (2016). Polymer/metal nanocomposites for biomedical applications. *Materials Science and Engineering: C, 60,* 195–203. https://doi.org/10.1016/j.msec.2015.11.023
6. Papageorgiou, G., Achilias, D., Nianias, N., Trikalitis, P., & Bikiaris, D. (2013). Effect of the type of nano-filler on the crystallization and mechanical properties of syndiotactic poly-styrene based nanocomposites. *Thermochimica Acta, 565,* 82–94.
7. Perna, A. S., Viscusi, A., Gatta, R. D., & Astarita, A. (2022). Integrating 3D printing of polymer matrix composites and metal additive layer manufacturing: Surface metallization of 3D printed composite panels through cold spray deposition of aluminium particles. *International Journal of Material Forming, 15*(2). https://doi.org/10.1007/s12289-022-01665-9
8. Kershner, D. (2022, December 8). The metal vs polymer 3D printed mould tooling debate: Which is "better?" *Fortify.* https://3dfortify.com/metal-vs-polymer-3d-printed-tooling/
9. https://printingit3d.com/how-to-resize-an-stl-file-the-complete-guide/
10. https://forum.bobcad.com/t/generate-stl-file-for-job-when-posting/284
11. Dizon, J. R. C., Gache, C. C. L., Cascolan, H. M. S., Cancino, L. T., & Advincula, R. C. (2021). Post-processing of 3D-printed polymers. *Technologies, 9,* 61. https://doi.org/10.3390/technologies9030061
12. Khosravani, M., Schüürmann, J., Berto, F., & Reinicke, T. (2021). On the post-processing of 3D-printed ABS parts. *Polymers, 13,* 1559. https://doi.org/10.3390/polym13101559.
13. Jaksic, N. (2015). *What to do when 3D printers go wrong: Laboratory experiences* (p. 122). ASEE Annual Conference and Exposition, Conference Proceedings.
14. www.fabbaloo.com/2015/10/is-your-3d-print-failing-we-found-a-troubleshooting-guide-for-you
15. www.scan.co.uk/buying-guides/3d-printers

6 Scopes for Using Waste Plastics

6.1. INTRODUCTION

Around a century ago, plastics began to appear in our daily lives. Today, they are still essential materials with a wide range of uses, whether people are using them at home, at work, when travelling, or in their free time. The applications for plastics are almost endless due to their extraordinary versatility as a material. Their advantages include without a doubt their great mechanical strength, low density, light weight, simple production, and affordable price (Mwanza and Mbohwa 2017). These characteristics have led to the use of plastics in a variety of industries, including the manufacturing of packaging, the automobile industry, electricity, construction, and transportation, as well as in fields like medicine and agriculture. The handling of the massive volumes of garbage produced by the pervasive plastic is a major issue. In 2018, there were 359 million metric tonnes of plastics produced worldwide (61.8 million metric tonnes in the EU) [1, 2]. Over the next two decades, it is expected that this number will increase by double. Plastic garbage is not handled and ends up in landfills in many nations. The amount of plastic that has to be kept is increasing quickly each year, and landfill space is scarce. To fulfil the needs of the circular economy project, stricter waste management standards must be enforced in order to ensure material and energy recovery. In the EU, 75.1% of plastic garbage was processed (32.5% recycled and 42.6% recovered energy), yet 24.9% was still dumped in the ground [2]. Managing the increasing volume of plastic trash may be done in a number of ways, such as basic recycling (re-extrusion), mechanical recycling, chemical repurposing, or using thermal technologies that produce energy (combustion, pyrolysis, gasification) (Al-Salem et al. 2009) [1].

Due to the COVID-19 pandemic's imminent emergence, the entire planet is in danger. The international community has implemented preventative measures to limit its transmission and spread due to its high contagiousness. The usage of face masks is one of the efficient strategies used by healthcare professionals and the general public across the world. Some nations embraced the use of facemasks early, while others did so later, to stop the spread of the COVID-19 virus. The World Health Organization (WHO) included the use of facemasks in its recommendations for preventing the virus' transmission in public settings (Worby and Chang, 2020). The rate of mask manufacture on a worldwide scale has increased significantly and will do so in the years to come. For instance, China is the world's top manufacturer of surgical masks. In February 2020, China produced 116 million face masks per day, which is 12 times more than typical (Adyel, 2020). Only 1% of face masks are managed properly, which might result in daily waste of 30,000–40,000 kg (World Wildlife Fund, 2020). In addition, these face masks contribute to the pollution of

DOI: 10.1201/9781003349341-6

microplastics by degrading into smaller particles (less than 5 mm) when exposed to the environment. (Zambrano-Monserrate et al., 2020) [3].

6.2. THE PROBLEM OF PLASTICS

Plastic is a polymeric substance, which means that it has very massive molecules that frequently resemble lengthy chains comprised of an apparently unlimited number of interconnecting connections. Although there are many natural polymers, including silk and rubber, they have not been linked to environmental contamination since they degrade quickly in the environment. Today, however, the typical consumer interacts on a daily basis with a wide range of plastic materials that have been created specifically to thwart natural decay processes. These materials, which are primarily made up of petroleum and can be moulded, cast, spun, or tried to apply as a coating, arise in a variety of shapes and sizes. Because they are essentially non-biodegradable in nature, manmade plastics tend to linger in natural settings. Additionally, a lot of lightweight single-use plastic items and packaging—which make up about 50% of all plastic products not placed in containers to be later sent to landfills, recycling facilities, or waste incineration. Instead, they are carelessly discarded at or close to the spot where they no longer serve the needs of the customer. They start to harm the environment as soon as they are dropped on the ground, flung out of a car window, piled atop a garbage can that is already full, or unintentionally swept off by a strong wind gust. In many places of the world, plastic packaging has become a widespread hazard in the landscape. (Illegal plastic waste disposal and breaching containment constructions also contribute.) Although metro areas produce the most trash, studies from throughout the world have not identified a single nation or demographic group as being the most guilty. Plastic pollution affects people everywhere, both directly and indirectly.

The ocean receives a large portion of the plastic debris produced on land since it is located downstream from almost every point on Earth. Each year, several millions of tonnes of debris—much of it carelessly dumped plastic litter—find their way into the world's seas. 2014 saw the release of the initial oceanographic investigation of the quantity of near-surface plastic trash in the world's seas. The amount of plastic floating on or close to the surface was calculated to be at least 5.25 trillion individual particles, each weighing around 244,000 metric tonnes (269,000 short tonnes). According to 2021 research, bags, bottles, and other things used for takeaway meals make up 44% of the plastic waste found in rivers, seas, and along shorelines. Oceans and beaches continue to get the majority of the focus of those researching and trying to reduce plastic pollution. The earliest evidence of ocean pollution with plastic came from scientists examining plankton in the late 1960s and early 1970s. Five subtropical gyres, which make up 40% of the world's seas, have been shown to amass floating plastic debris. These gyres, which are midlatitude on Earth, include the North and South Pacific Subtropical Gyres, whose eastern "garbage patches" (regions with large quantities of plastic debris circulating near the ocean surface) have caught the attention of scientists and the public. The North and South Atlantic Subtropical Gyres, as well as the Indian Ocean Subtropical Gyre, are the remaining gyres. Plastic pollution in the waters can destroy sea life directly by entangling

FIGURE 6.1   Plastic Pollution in Our Country [5].

them in items like fishing tackle, but it can also kill by being mistaken for food and being ingested by animals. Numerous species, especially tiny zooplankton, huge cetaceans, the majority of seabirds, and all marine tortoises have been discovered to readily consume plastic fragments and garbage objects like cigarette lighters, plastic bags, and bottle caps, according to studies. Plastic becomes brittle when exposed to sunlight and saltwater, and it becomes available to zooplankton and other small marine organisms when bigger materials eventually break down into microplastic particles. Only about 5 mm (0.2 inch) long fragments of plastic make up a significant portion of the plastic debris found in the seas. More than 114 aquatic species, including those that can only be found in the deepest ocean abyss, had microplastics discovered in their organs by the year 2018. Researchers have found that deep-sea currents are causing microplastic "hot spots" in some areas of the seawater, such as one in the Tyrrhenian Sea that already had nearly 2 million microplastic pieces per square metre. By 2020, scientists predicted that at least 14 million metric tonnes (15.4 million short tonnes) of microplastic particles would be taking it easy on the ocean floor (about 186,000 pieces per square foot) [3, 4].

6.3.   POLLUTION BY PLASTICS ADDITIVES

In particular, the release of chemicals employed in its manufacturing causes plastic to contaminate even when it is not littered. In fact, one rising area of worry is the contamination of the environment by chemicals released into the air and water by plastics. This has led to increased examination and regulation of several chemicals used in plastics, including phthalates, bisphenol A (BPA), and polybrominated diphenyl ether (PBDE). Phthalates are plasticizers—softeners used to reduce the brittleness of plastic goods. Along with medications, colognes, and cosmetics, they can be found in electronics, packaging materials, automotive upholstery, carpet materials, and medical gadgets. BPA can be found in packaging, bottles, CDs, medical gadgets, and the interior of food cans since it is used to make strong epoxy coatings and adhesives and transparent, durable polycarbonate plastic materials. Plastics are given PBDE as a flame retardant. These substances have all been found in people and are all known to interfere with the endocrine system. BPA resembles the naturally occurring female hormone oestrogen, PBDE has also been demonstrated to affect thyroid hormones, and phthalates are classified as anti-androgens because they interfere with male hormones. Children and women who are close to menstruating are the groups most susceptible to these hormone-disrupting substances.

Animals living in terrestrial, aquatic, and marine settings have also been linked to hormone disturbance by these substances. Blood levels that are lower than those observed in the typical inhabitant of a developed nation cause effects in experimental animals. Fish, mollusks, worms, invertebrates, crustaceans, and amphibians all exhibit effects on their life processes, including changes in the number of offspring born, disruptions in larval development, and (in the case of insects) delayed emergence. Studies examining population declines as a result of these effects have not been documented, however. Investigations are required to close this information gap, as well as studies on how exposure to mixes of those substances affects both people and animals.

6.4. PLASTIC POLLUTION IS A HUMAN HEALTH ISSUE

The use of plastic is all around us. It's used in a lot of our furniture, clothes, equipment, and packaged food. Natural materials used in production, such as paper, glass, and cotton, have been supplanted by plastic during the previous few decades. We are aware that the tremendous plastic contamination of our environment is a result of the pervasive usage of plastics. Plastics, meanwhile, affect more than simply the environment. As pointed out by toxicologist Prof. Dr. Dick Vethaak, "we are dealing with a human health issue as well."

Plastics may affect our health via three pathways:

1. We eat, drink, and breathe microplastics every day. These small plastic particles may harm our health once they have entered our bodies.
2. Plastic products contain chemical additives. A number of these chemicals have been associated with serious health problems such as hormone-related cancers, infertility, and neurodevelopment disorders like ADHD and autism.
3. When plastics and microplastics end up in the environment, they attract micro-organisms, such as harmful bacteria (pathogens). If microplastics containing these pathogens enter our body, they may increase the risk of infection [6].

6.4.1. MICROPLASTICS AND PATHOGENS

Microplastics are plastic granules with a diameter of 5 millimetres or less. Nanoplastics are polymers that are even a million times smaller than a millimetre. Everywhere in our environment, even the deepest ocean tunnels, both poles, and the summit of the Himalayas, microplastics have been discovered. Microplastics can also be discovered considerably closer to home, though. Microplastics are pervasive in almost every home since they are emitted from our couches, carpets, curtains, and other synthetic materials. Numerous consumer goods, including seafood, honey, wine, water, salt, fruits, and vegetables, have also been discovered to contain microplastics.

We eat, drink, and inhale plastics each day as a result of plastic pollution of our environments both indoors and out. Because plastics cannot biodegrade, microorganisms that are known to cause human illnesses frequently flourish on their surface. This begs the urgent question: Do microplastics make us ill once they're inside of us? A unique programme called Microplastics & Health was started by ZonMw, a Dutch organisation that funds health-related research, to uncover the answers. The Netherlands is establishing itself as a leader in this subject on a worldwide scale with these research initiatives. The first field investigation of how microplastic contamination may impact soil fauna was published in the Journal of the Royal Society in 2020. The study finds that species that exist underneath the surface, like mites, larvae, and other microscopic organisms that keep the soil fertile, have declined as a result of terrestrial microplastic contamination.

Chlorinated plastic has the potential to leak dangerous substances into the soil around it. These chemicals can then seep into nearby water sources, including aquifers, and the environment. The animals that consume the water may experience a variety of potentially dangerous impacts as a result [7].

6.4.2. CHEMICAL ADDITIVES

Plastics are given certain properties by adding chemicals, such as plasticity, colour, malleability, durability, or the hardness required by some items. Some of these chemical groups, however, have been recognised as potentially harmful to human health because, for instance, they may interfere with our hormonal system. Bisphenol A (BPA), an endocrine-disrupting chemical (EDC), is found in plastic boxes, canned-food liners, polycarbonate bottles, and more. Since hormones are crucial to the development of both foetuses and children, these groups are thought to be particularly vulnerable. A disturbing fact was *shared* by endocrinologist Prof. Dr. Laura Vandenberg: "Every child born on this planet, is born pre-polluted. With dozens, or perhaps hundreds of chemicals that are found inside of their bodies." The dangerous compounds that have been found have indeed been linked to a number of health difficulties, including poor neurodevelopment in kids, immunological diseases, and an increased risk of "hormone-related malignancies" [4].

6.5. HOW TO TURN PLASTIC WASTE INTO 3D PRINTING PROFIT

Researches in the field of sustainable design have recently made design for the circular economy their main approach (Sauerwein et al. 2019). A progressive taxation

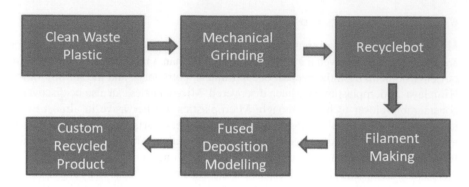

FIGURE 6.2 DRAM Process.

system between recovery operations provides the stabilization of the material, or the degree to which the material remains the same from its uniqueness, while the key components of this approach are the extension of material life and total recovery of components and its architecture (Hollander et al. 2017). In other words, in the material circular economy, waste is kept to a minimum and product value is preserved for as long as feasible (European Commission 2015) [8].

If recycling can earn them money, individuals will do it. Metal and glass recycling have been quite successful in areas where cans and bottles can be exchanged for cash. Sadly, there haven't been as many incentives to recycle plastic. Only 9% of plastic garbage is recycled as of 2015. The remainder contaminates the environment or landfills. However, as new technologies have developed, individuals may now directly recycle plastic waste by 3D printing it into useful things at a fraction of the expense of doing so. Making toys and games, recreational equipment, electronics, accessories, and home and yard items from recycled material from millions of free designs is a growing trend. Distributed recycling and additive manufacturing, or DRAM for short, is the name of this strategy.

Sorting and washing the plastic with soap and water or even putting it through the dishwasher is the first step. The plastic must next be crushed into tiny pieces. A crosscut paper/CD shredder performs well for modest quantities. Online resources for open-source industrial waste plastic granulator designs exist for bigger quantities. You then have a few options. A recyclebot, a machine that transforms pulverised plastic into the spaghetti-like filaments used by the majority of low-cost 3D printers, may convert the particle into 3D printer filament. Comparing the price of filament produced using a 3D-printable recyclebot to conventional filament, which may cost up to US$10 per pound, the latter is extraordinarily inexpensive, costing less than a cent per pound. Making commodities at home from garbage is more alluring now that the epidemic is disrupting global supply systems.

The second approach is newer: Fused particle manufacturing allows you to 3D print granulated plastic wastes into goods without having to first create filament. This method may be used with desktop printers as well as bigger printers like the open-source, for-profit GigabotX printer, which are best suited for printing massive goods.

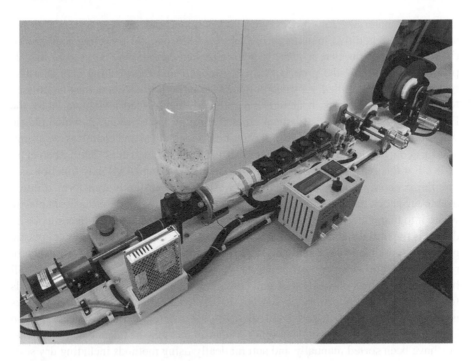

FIGURE 6.3 Recyclebot Technique [9].

A syringe printer may also be used to print directly on granulated plastic debris, however this method is less common because it requires refilling the syringe after each print. The two most common polymers for 3D printing, ABS and PLA, as well as a large list of plastics you almost certainly use every day, such as PET containers, have been proved to function with this technology. Any plastic garbage with a recycling emblem on it may now be turned into useful items. DRAM can even function in remote regions without recycling or electricity thanks to an "ecoprinting" project in Australia that uses solar power. As a result, DRAM is applicable almost wherever there are people, a lot of plastic debris, and sunlight. Custom items may be created using DRAM for less money than the sales tax on typical consumer goods. There are currently millions of free 3D-printable designs available for everything from home items to adaptive devices for those with arthritis. These things are already being 3D printed by prosumers, who are together saving a lot of money.

6.6. POLYMER RECYCLING IN AM—AN OPPORTUNITY FOR THE CIRCULAR ECONOMY

It is important to take precautions during any manufacturing process to reduce waste production and harm to the environment. The kind of raw materials used in the manufacturing process and the way waste is disposed of are crucial factors in determining the environmental effect. The FDM technique may utilise recyclable thermoplastics as its basic materials. Any process' residual polymeric waste may be recycled and utilised in the additive manufacturing process. Mechanical recycling and chemical recycling are the two primary recycling processes that may be used to recycle plastics. However, if the properties are optimised and the printing characteristic requirements are fulfilled, recycled polymers from both procedures may be utilised to produce AM feedstock. It should be mentioned that the feedstock material for the FDM-based AM method is filament created through the extruder, in which plastic waste may be utilised directly (or just after both mechanical and chemical recycling) and turned into filaments (Herianto et al. 2019; Rahimizadeh et al. 2019; Zhao et al. 2018). For this reason, FDM-based additive manufacturing (AM) may function as a circular economy approach where plastic waste can be instantly transformed into a usable feedstock material. Figure 6.5 depicts the procedure used to transform plastic waste into filament, which is then processed into 3D printed items. Collecting plastic debris is the initial stage in the recycling of plastic waste. Direct collection of plastic garbage is possible from businesses, large merchants, and municipal governments. In recent times, curbside drop-off, repurchase, and deposit/refund procedures have all been used to collect plastic bottles and containers from the general public (Methods for Plastic Wastes Collection | EcoMENA 2020). After collecting, separating plastics is a crucial step in which garbage is divided into different categories according on kind and grade, and any metals or other contaminants may be taken out. Waste plastics have been sorted manually and automatically using methods including dry sorting, air sorting, electrostatic sorting, and mechanical sorting (Rozenstein et al. 2017). To remove any foreign elements, the sorted plastic wastes are cleaned, chopped into bits, and sorted again. After that, the waste is extruded to create filaments for the FDM process by adding stabilisers, plasticizers, and lubricants as needed.

6.7. POLYMERS/COMPOSITES MADE FROM RECYCLED MATERIALS THROUGH FDM

The possibility of developing a large-scale production method based on FDM would be made possible by the rapidity and flexibility with which complicated designs might be manufactured. The utilisation of recycled trash in FDM is required because of the aforementioned. Marciniak et al. (2019) utilised waste ABS material from the printer to make specimens using the FDM technique. Tensile strength, flexural strength, and impact strength measurements for the recycled ABS specimen were 41 MPa, 61 MPa, and 30 kJ/m^2, respectively. Due to polymeric breakdown during regeneration, recycled ABS showed a 13–49% loss in tensile strength by Iqbal et al. (2017). However, this could be reduced by optimising the printing parameters. Cruz et al. (2017) presented a technique for recycling plastics in FDM and investigated the use of recovered PLA in additive manufacturing. The investigation's findings showed that recycled PLA could be used to create AM samples, but further recycling wasn't really advised since the polymer would degrade. In another work, Alberto et al. (2015) observed, due to frequent recycling, a decrease in the ultimate tensile rate of the PLA polymer. After five recycling cycles, the PLA showed a 10% decrease in elongation at break, but an improvement in tensile strength and modulus. Elsewhere,

Babagowda et al. (2018) produced polymer composites based on FDM using recycled PLA. In order to create composites, recycled PLA was mixed with virgin PLA at different ratios (10, 20, 30, 40, and 50%). The recycled PLA sample showed good mechanical characteristics, with the 10% recycled PLA-added composite having the highest tensile strength (38 MPa) and flexural strength (4 MPa). For the creation of a recycled PLA sample, the printing settings needed to be optimised. The outcome recommended decreasing the layer height to attain better mechanical properties. Mägi et al. (2015) employed recycled polyamide 12 (PA12) for printing composites made with recycled PA12 polymer. As a byproduct of selective laser sintering, PA12 was obtained. When opposed to virgin PA12, the recycled PA12 had a 6% lesser tensile strength. However, when discarded PA12 was mixed with 10% polyurethane [10], the tensile strength was improved by 12%. The composite produced by reinforcing recycled PA12 with 5 weight percent aramid fibre performed similarly to that of virgin PA12. The results of the experiment suggested that the recycled polymer's fibre reinforcing might boost its strength. Mägi et al. (2016) studied, through the development of recycled PA12 composites with various weight ratios of graphite particles (10, 20, 30, and 40 wt%), the influence of particle reinforcement inside the recycled PA12. Although equivalent to virgin PA at 30 weight percent, the tensile strength of such graphite/recycled PA12 composites was lesser compared to virgin PA12 (39 MPa). Singh and Singh (2016) created a composite utilising recycled polyamide 6 (PA6) with reinforcement from aluminium and aluminium oxide. After being recycled and fortified, the PA6 was extruded into filaments. A PA6, 30 wt% Al, and 10 wt% Al2O3 FDM composite was created. The wearability of the composite was evaluated and contrasted with that of the commercial material. When opposed to the commercial material, the recovered PA6 composite showed a reduced rate of degradation [8].

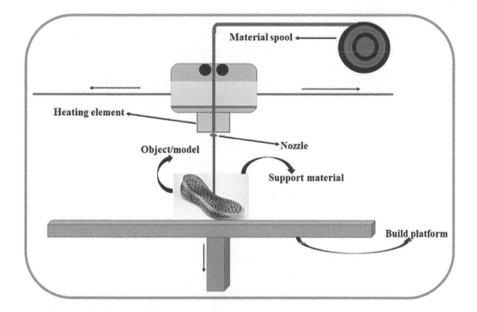

FIGURE 6.4   FDM 3D Printing.

6.7.1. POTENTIAL AND SCALE UP

Even while rapid prototyping and the quick and simple creation of innovative objects are made possible by 3D printing, it nonetheless adds to the rising quantity of easily accessible consumer goods that are most likely to end up in landfills.

The procedure is less wasteful than some other production techniques and conforms more closely to the 3 Rs, but there is still post-processing waste and failed components to take into account. To avoid distortion when printing, 3D-printed objects frequently require supports; how these are disposed of must be taken into account when evaluating how sustainable the procedure is [11]. No chance that 3D printing will become the dominant means of production any time soon. Even while rapid prototyping and the quick and simple creation of innovative objects are made possible by 3D printing, it nonetheless adds to the rising quantity of easily accessible consumer goods that are most likely to end up in landfills.

6.7.2. BENEFITING FROM THE POPULARITY OF CUSTOMISED PRODUCTS

The popularity of 3D printing in industries like hearing aids demonstrates that this method of manufacturing has a competitive advantage when producing goods with intricate shapes. It demonstrates that 3D printing is a powerful tool for creating goods that are uniquely tailored. The expenses of modifying the product are quite minimal, in contrast to conventional production techniques. Products that may be customised include jewellery, eyeglasses, and footwear. These are lucrative marketplaces for businesses that use 3D printing to create goods for other people.

6.7.3. UNTAPPED POTENTIAL

Industry analysts predict that 3D printing might expand dramatically in the years to come, despite the slim probability that it would ever overtake other production techniques in manufacturing.

According to Terry Wohlers, industry turnover will reach $115 billion in 2030 with an average annual growth rate of 27%. That would represent over 1% of global manufacturing and be nine times the size of the sector as it is today. Wohlers anticipates that 3D printing will eventually be able to account for at least 5% of manufacturing. It will take around 20 years to get there if the annual rate of growth of 27% is maintained (2040). Although the 27% growth prediction is slightly higher than that of the 25% average growth in the three years prior to the crisis, we still believe that 27% is a realistic growth rate for a number of reasons:

- **Threshold fear diminishing**

Growth in 3D printing has always been hampered by threshold anxiety. Since 3D printing has shown itself throughout the pandemic, saving the day amid supply shortages, we anticipate this to decrease dramatically. As we already said, when it comes to incorporating 3D printers into their manufacturing process, consumers are currently in a "rapid forward" frame of mind, according to consultants. We believe that this trend may potentially have an upward impact on 3D printing investment of more than two percentage points.

- **Lack of suitable materials is being overcome**

In the past, the shortage of suitable materials has been a bottleneck, but specialists in the field claim that advancements in building and polymers are resolving this issue. There will soon be a broader variety of materials, which will encourage more businesses to use the new technology.

- **New adopters are entering the market**

We anticipate that as more and more new businesses and occasionally entire industries begin using 3D printers, the proportion of 3D printed goods in all manufactured goods will continue to rise. For instance, the US military has begun using 3D printers to create missile launcher shelters. In recent times, 3D printers have also been embraced by the power and energy industries. There is absolutely no reason to believe that this will abruptly come to an end. According to the Wohlers Report (2021), there were 228 suppliers of industrial 3D printers as a whole in 2020. This is a sevenfold increase from 2012.

According to the same research, the top eight suppliers still had a combined market share of 14% just three years ago, but that number dropped to 9% last year. In accordance with this, 2020 also revealed a loss in the major eight providers' turnover while smaller suppliers produced a rise in income. We see the growth in suppliers as a sign of more competition, which typically leads to more downward pressure on prices. This is good news for the economy for 3D printing since it will increase demand.

- **It's all about adaptability**

The growing popularity of customised items will be advantageous for 3D printing because it is much more affordable to make them than traditional machinery. One in six customers has at least once purchased a product that was customised.

We believe Wohlers' prediction of a future proportion of 5% of 3D printed items in overall production is reasonable in light of all of these factors. We estimate that it may be attained by the year 2040.

In our 2017 study on 3D printing, we examined the potential for 3D printed goods to account for between 25–50% of all produced goods. Only once mass production using 3D printing becomes economically viable will this scenario come to pass. However, there is currently no sign that such a breakthrough is imminent.

- **Reshoring is about more than economics**

The decision to reshore manufacturing is no longer only an economic one driven by cost-benefit calculations for businesses in the public discourse. Reshoring calls are bolstered by other movements that call for shifting from complex global value chains to more "local for local" production. These movements include criticism of the ability to contribute of cross-border distribution networks to climate change, effects on local workforce, infringements of workers' rights, and tax evasion.

- **Protectionism is a potential push for 3D printing**

If it gets more and more popular among politicians, forcing governments to alter the playing field, reshoring could become required for businesses. Raising tariffs or other trade obstacles might alter how affordable it is to produce in industrialised nations, forcing businesses to move their facilities back home or to nations with which they have free trade agreements.

Several businesses have relocated as a result of the increased tariffs the US placed on Chinese imports. According to the *Financial Times*, some of them travel back to the US while others go to nations like Vietnam and Thailand.

For the time being, the conclusion is that the majority of businesses do not believe that reshoring is the best solution to supply chain problems. But the more consumers learn about labor-saving industrial techniques like 3D printing, the more appealing it may be, particularly if politicians intensify protectionist measures [12–14].

Consequently, it will be crucial to monitor how reshoring initiatives are developing [15].

6.8. SCALE UP STUDIES

Since the introduction of 3D printing in the middle of the 1980s, additive manufacturing has expanded gradually and discovered several uses in a variety of sectors. In more recent times, when the initial patents' periods came to an end and new businesses joined the market, the sector had a growth spurt. There are many various additive manufacturing techniques, but polymer-based material extrusion 3D printing—also known as fused filament fabrication—has grown to be one of most popular due to its affordability, use, and variety. Despite significant advancements in print quality and availability of raw materials, physical sizing has frequently been a problem for material extrusion 3D printers. When the print begins, there are fundamental design challenges to the extrusion process.

There are problems with the mechanics built into the process when a material extrusion 3D printer is scaled up. When a result, various design modifications are required as the machine's horizontal construction area rises over 500 mm × 500 mm. The ideas outlined in Section 3.4 were used to derive the 500mm × 500 mm limitation. Beyond this scale, 3D printing restrictions caused by extrusion techniques become an issue.

6.8.1. FRAME DESIGN

The design of 3D printers is often either an open-air idea or a contained concept. Due to its straightforward construction and less expensive design, open-air designs are quite popular among hobbyists. Boxed designs, on the other hand, work well with the more expensive printers used in offices. Due to its straightforward design and straightforward fabrication, the open-air design is well liked among lower-priced 3D printers. Frames that completely or partially encircle the printer define boxed designs. Mechanisms are shielded from dust in such a design, requiring less maintenance. Additionally shielding the user from harm like burns or pinched fingers, boxed designs help further manage the environment in which the print is shown.

6.8.2. KINEMATIC CONFIGURATIONS

In order to produce the three-axis movement, a variety of different kinematic systems are utilised in material extrusion 3D printers. The employment of timing pulleys and motors to drive motion in the horizontal x- and y-axes, along with power screws to drive motion in the vertical z-axis, is the most common of these methods. This is so that the z-axis, which demands extremely high precision but moves seldom, can perform the high speed, high acceleration, and jerk movements required by the x- and y-axes. Power screws have been utilised in certain designs for all three axes of motion, although belt conveyors are more popular due to their faster motion and lower cost.

The mass of the component to be produced, particularly while the print bed is moving, represents one of the main problems when sizing 3D models. Large printed parts would have tremendous momentum, necessitating the use of a considerably more durable kinematic system. Therefore, a system with a fixed print bed is preferable. The vertical z-axis in this scenario would be independent and often powered by a power screw. There are three main primary movement methods for the x- and y-axes: "CoreXY" (also known as D-bot), "H-drive," and direct drive.
The CoreXY and H-drive setups' motion kinematics are as follows.

$$\Delta p = \begin{bmatrix} \Delta x \\ \Delta y \end{bmatrix} = rJ\Delta\theta \tag{1}$$

$$J = \frac{1}{2}\begin{bmatrix} 1 & 1 \\ 1 & -1 \end{bmatrix} \tag{2}$$

$$\Delta\theta = \begin{bmatrix} \Delta\theta_1 \\ \Delta\theta_2 \end{bmatrix} \tag{3}$$

Both CoreXY and H-drive configurations have the same equation of motions because they follow the same belt path. The belt is divided in half by the CoreXY, however. In order to shorten the belt and avoid applying torque to the carriage, this is performed [16]

6.8.3. CONTROLLING THE TEMPERATURE OF THE SURROUNDING ENVIRONMENT

Internal strains in the component brought on by temperature differences from freshly deposited material and the existence of the print nozzle are what lead to warping. One of the variables that has the greatest impact on the structure and binding between

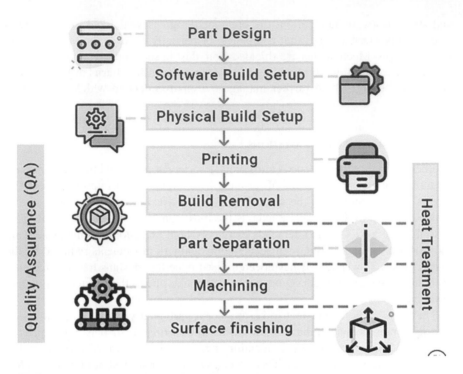

FIGURE 6.5   Scale Up Studies of 3D Printing [17].

each layer of the print is discovered to be the nozzle temperature [17]. As the print size grows, the temperature differential between the part's ends that is permitted to exist decreases. This has an inverse squared connection, thus as the part grows in size, the problem of warping brought on by thermal stress is much worse [10]. dT dzmax ¼ 8δ L2α δ5Þ, where L is the printed part, is actually horizontal width in millimetres, dT/dzmax is the greatest difference in temperature along the z-axis expressed in degrees Celsius per millimetre, and is the print material's thermal gradients coefficient. The surrounding air can be heated to slow down the cooling rate of the printing, which helps to lessen the problem of warping. Internal thermal tensions are lessened and can be gradually eased by lowering the cooling rate, hence reducing strain. A heated build plate is frequently included with 3D printers to help warm the environment around the print. Additionally, some printers include enclosures to stop draughts and balance out the temperature of the air around them. An open-air design with a heated bed works well for small printers, but larger machines have problems with this. The electricity needed to heat a bigger print surface will be substantially greater, increasing the machine's operating expenses.

6.8.4.   OVERALL PRINT TIME

The layer resolution of the print is one of the most important criteria that affects print time. One of the key determinants of print quality is layer resolution, which is the thickness of each layer of the print. In order to increase dimensional accuracy, it is ideal to have thin print layers. However, a thin layer necessitates the application of additional layers, which adds to the printing process' time need. It is possible to estimate the print time at which the material is actively laid for a print with about consistent cross-sectional area as described in the following: Original Layer Thickness * Print Time/Average Time per Layer Altered Layer Thickness. The increment in print time might become extremely noticeable for bigger prints. To balance the trade-off between print quality through layer thickness and overall print time, new techniques must be investigated. Some techniques, like the Adaptive Layers tool in Ultimaker Cura and the Custom Layer Smoothing feature in Slic3r Prusa Edition, were previously investigated.

6.9. THE FUTURE OF 3D PRINTING IN MANUFACTURING

Future industrial production will probably combine conventional techniques with 3D printing. The latter has a tendency to be more resource-efficient and allows for more innovative and efficient designs, which helps to reduce the amount of production needed. Compared to conventional manufacturing methods, it is also cleaner, greener, and more viable, although there is still potential for development.

REFERENCES

1. Mikula, K., Skrzypczak, D., Izydorczyk, G., Warchoł, J., Moustakas, K., Chojnacka, K., & Witek-Krowiak, A. (2020). 3D printing filament as a second life of waste plastics—a review. *Environmental Science and Pollution Research*, *28*(10), 12321–12333. https://doi.org/10.1007/s11356-020-10657-8

2. www.plasticeurope.org, accessed April 9, 2020.

3. Chowdhury, H., Chowdhury, T., & Sait, S. M. (2021). Estimating marine plastic pollution from COVID-19 face masks in coastal regions. *Marine Pollution Bulletin*, *168*, 112419. https://doi.org/10.1016/j.marpolbul.2021.112419

4. Kurtela, A., & Antolovic, N. (2019). The problem of plastic waste and microplastics in the seas and oceans: Impact on marine organisms. *Croatian Journal of Fisheries*, *77*, 51–56. https://doi.org/10.1016/10.2478/cjf-2019–0005.

5. www.indiatimes.com/technology/science-and-future/10-tech-solutions-plastic-pollution-544071.html

6. How Plastic Affects & Threatens Human Health. (2022, March 2). *Plastic health coalition*. www.plastichealthcoalition.org/

7. United Nations Environment Programme. (n.d.). Plastic planet: How tiny plastic particles are polluting our soil. *UNEP*. www.unep.org/news-and-stories/story/plastic-planet-how-tiny-plastic-particles-are-polluting-our-soil

8. Shanmugam, V., Das, O., Neisiany, R. E., Babu, K., Singh, S., Hedenqvist, M. S., Berto, F., & Ramakrishna, S. (2020). Polymer recycling in additive manufacturing: An opportunity for the circular economy. *Materials Circular Economy*, *2*(1). https://doi.org/10.1007/s42824-020-00012-0

9 Pearce, J. (2018, May 27). RepRapable recyclebot [Video]. *YouTube*. www.youtube.com/watch?v=b04mUaI-oTU

10. Shanmugam, V., Das, O., Neisiany, R. E., Babu, K., Singh, S., Hedenqvist, M. S., Berto, F., & Ramakrishna, S. (2020). Polymer recycling in additive manufacturing: An opportunity for the circular economy. *Materials Circular Economy*, *2*, 11. https://doi.org/10.1007/s42824-020-00012-0

11. Taylor-Smith, K. (2021, January 12). How is 3D printing a sustainable manufacturing method? *AZoM.com*. www.azom.com/article.aspx?ArticleID=20017

12. Reichental, A. (2020). When it comes to 3D printing, how much sustainability is enough? *tct mag*. Retrieved January 7, 2021, from www.tctmagazine.com/blogs/guest-column/3d-printing-how-much-sustainability-is-enough

13. AMFG (2020). How sustainable is industrial 3D printing? *AMFG*. Retrieved January 7, 2021, from https://amfg.ai/2020/03/10/how-sustainable-is-industrial-3d-printing/

14. Folk, E. (2019). How sustainable is 3D printing? *3DPrint.com*. Retrieved January 8, 2021, from https://3dprint.com/239063/dhow-sustainable-is-3d-printing/

15. Leering, R. (2021, August 9). 3D printing's potential. *ING Think*. https://think.ing.com/articles/3d-printing-potential/

16. Idà, E., Nanetti, F., & Mottola, G. (2022). An alternative parallel mechanism for horizontal positioning of a nozzle in an FDM 3D printer. *Machines*, *10*(7), 542. https://doi.org/10.3390/machines10070542

17. How to Scale Additive Manufacturing: From Concept to Mass Production. (n.d.). https://globalluxsoft.com/how-to-scale-additive-manufacturing-from-concept-to-mass-production

7 3D Printing vs Injection Moulding

7.1. BASICS OF INJECTION MOULDING

Injection Moulding (IM) is a fast and dynamic manufacturing method used in the plastics industry to develop objects of various sizes, shapes, and, if necessary, many details. It entails injecting melted thermoplastic or thermoset materials into a closed mould at high pressure and temperature. At the end of the manufacturing cycle, the finished product cools and/or solidifies inside the mould before being ejected. When heated, a thermoplastic polymer needs to undergo a physical change, a transition to a viscous state, allowing it to be moulded into the desired shape. When reworking is required, this process can be repeated several times (e.g. polyethylene, PE; polyvinyl chloride, PVC). A thermoset polymer, on the other hand, solidifies as a result of heat-induced cross-linking. This chemical modification is thus responsible for the resulting product's hardness. Because reheating would cause degradation, the latter cannot be moulded again (e.g. phenol formaldehyde resins like bakelite) [1]. IM is a relatively new technique, having emerged between the end of the 1800s and the beginning of the 1900s, with a real explosion around 1940 associated with increased demand for low-cost goods. The advancement brought about by IM in plastics processing over previous manufacturing techniques is based on the simultaneous use of pressure and heat to transform a polymer, or a polymeric formulation, into a solid object with defined shape, dimension, and features [2].

7.1.1. PROCESS AND APPLICATIONS

Injection moulding is ideal for mass-producing plastic parts with complex shapes that require precise dimensions. The hot polymer melt is forced into a cold empty cavity of the desired shape and allowed to solidify under high holding pressure in this process. Filling, post-filling, and mould-opening are the three stages of the injection moulding cycle. During the moulding process, the plastic material experiences temperature and pressure increases, significant shear deformation, and rapid decay of temperature and pressure in the mould cavity, resulting in solidification and locking of residual stress, orientation, and other part properties that determine the moulded part quality.

The IM process is carried out in appropriate equipment, IM machines, which typically have two parts: the plasticating/injecting unit (PIU) and the clamping unit (CU). Horizontal, vertical, and hybrid IM machines are distinguished based on their configuration. The latter displays horizontal PIU and vertical CU, or vice versa. PIU is made up of a hopper that feeds a heated barrel that goes through heating, mixing,

DOI: 10.1201/9781003349341-7

compression, and melting steps. Temperatures can be set and maintained using different heater bands located along the barrel. The barrel contains a pressure-creating element. Historically, this was a plunger (Hyatt's IM machine, 1872), but modern equipment is screw-type. Moving forward, the screw acts like a plunger, exerting the necessary injection pressure, whereas rotating results in the movement/compression of the solid material/melt, as well as the development of shear forces, which aid in the increase of the latter's temperature (mechanical heating). Despite the fact that the screw design should be tailored to the properties of the processing material, the metering screw is the most common. The rear section (feed zone) of such a screw has a smaller diameter than the front end (metre zone), where the material is forced to flow into a progressively narrower space, resulting in increased speed and frictional heat (squeezing action). The screw's intermediate zone (melt zone) serves as a transition zone between the metre and feed zones [4].

The screw is finished with a tip that fits the nozzle cap. A non-return valve located before the tip prevents the melt from flowing backward. The nozzle area is even smaller, as it is heated by its own heater band. Depending on the thermal stability and viscosity characteristics of the material, unwanted temperature increases may occur, potentially resulting in degradation phenomena. The mould is the IM machine's final component. It is typically made up of two halves that come together to form a cavity with a defined 3D shape that forms the moulded object's outer surfaces (single unit production cycle). It is also possible to develop moulds with multiple cavities in order to produce more than one unit within the same cycle. One half of the mould is fixed to a platen, while the other is mobile, allowing the two halves to be matched (closed mould) or uncoupled (open mould). During injection, the clamping unit keeps the mould closed. To keep the cavity pressure and prevent the mould from opening while the substrate is being injected, a clamping force greater than the injection pressure is required. The resulting gap would certainly impede the completion of the moulded objects and/or cause the melt to squeeze out of the mould cavity (short shot and/or flash). The temperature of the mould is controlled by a cooling system that typically uses water as the circulating fluid. After injection, the melt cools and/or solidifies in the mould, and when sufficiently hardened, it can be released by pins located in the mould's mobile half. However, in order to achieve its final mechanical properties, the moulded object may require a curing treatment. The object's size can vary in relation to the cavity image. This one is understood as shrinkage, and it may happen within the mould or even after ejection (mould and post-mould shrinkage, respectively). Warpage, on the other hand, is a deformation that involves the bending or twisting of the unit, resulting in changes in bi- and tridimensional shape; it is a type of non-uniform shrinkage. IM machines are either single-stage, with plastication and injection performed in the same cylinder by a reciprocating screw, or two-stage, with a plasticating screw feeding the melt into a holding/accumulator chamber [4]. The melt is injected into the mould cavity during the second stage (ram injection). The mould's filling/opening/closing cycle is coordinated with the screw movement (that moves forward to the injection position and is pushed back to the pre-injection one). The IM cycle restarts when the moulded object is automatically ejected, resulting in a continuous manufacturing process[1, 2].

Injection moulding is one of the most important processes for mass production of thermoplastic objects, typically without the need for additional finishing. Most injection moulding machines today are universal, which means they can accept all types of moulds within certain limits. This process has superior economics for articles with complex geometry, giving it an advantage over other techniques. Despite the high initial cost of injection moulding machines, cost per moulding decreases with scale. The injection moulding principle is very simple. The plastic is heated until it forms a viscous melt. It is then pressed into a closed mould, which defines the shape of the finished product. The material is cooled until it solidifies, at which point the mould is opened and the finished part is extracted. Although the principle is straightforward, the practice of injection moulding is anything but. This is due to the complex behaviour of plastic melts and the process's ability to encompass complex products. Heat transfer and pressure flow are critical mechanisms in injection moulding. An injection moulding machine, also known as a press, and a mould, which may also be referred to as a tool or a die, are required. The process produces moulding, which is sometimes referred to incorrectly as a mould.

Injection moulding is a process that involves heating a thermoplastic polymer above its melting point, which results in the conversion of the solid polymer to a molten fluid with a low viscosity. This melt is mechanically forced, or injected, into a mould in the desired final shape. Because of the low viscosity of the molten polymer, the article can be completely filled into the mould until it is cooled below the polymer's freezing point. In the case of semicrystalline polymers, the object's crystallinity (which governs its mechanical and appearance properties) is usually controlled by cooling the object in-mould at a predetermined rate. Finally, the mould is opened and the part is ejected and recovered [4].

The injection moulding process is divided into two stages: filling and holding. The screw moves the plasticized material forward at a constant speed during the plasticizing stage of the injection process. The material is compacted and continuously conveyed forward as the screw groove becomes shallower. The head is constantly accumulating while waiting for the injection instruction, and the screw is constantly retreating as the injection machine's back pressure increases during the injection process. When the injection is started, the screw advances and the material continuously fills the mould. At the same time, as the pressure in the mould rises, the screw moves while injecting. When the material is filled into the mould, the injection machine uses back pressure to inject the material into the mould. As the temperature of the material continues to fall, the pressure in the mould cavity begins to fall. When the injected material can be safely moulded without being damaged, the injection mould is opened and the product is ejected through the mould structure—products that can be inlaid with the mould before the mould is closed for the next cycle [5–8]. The injection product's external structure and size are identical to the internal mould of the injection machine, resulting in a mould-shaped product. Many defects occur in injection moulded products, including short shots, sprays, sags, flow marks, weld marks, and floating fibres. These defects can be treated by subsequent processes such as spray coating, but the cost and time involved will increase. Moulding with rapid heating and cooling cycles has proven to be an effective solution to this problem. At the same time, one of the most effective methods of producing high-quality precision injection products is to control the parameters in the injection moulding process.

7.2. INJECTION MOULDING EQUIPMENT, MOULDS, AND PROCESS

Because of its light weight, stable chemical properties, good insulation properties, and low price, plastic has gradually replaced many metal materials and is widely used in high-end industries and daily life. The term "responsibility" refers to the act of determining whether or not a person is responsible for his or her own actions (steel, wood, cement). They are widely used in the fields of agriculture, machinery, medicine, aerospace, automotive, and construction. The plastics industry is rapidly expanding. The injection moulding process can form a variety of complex shapes, sizes, and precise plastic products. It has short cycle times, high efficiency, and high precision.

7.2.1. INTRODUCTION OF INJECTION MOULDING MACHINE

The injection moulding machine, also known as the plastic injection moulding machine or injection machine, is used to add granular or powdery polymer raw materials to the barrel of the injection moulding machine, melt and plasticize into polymers with good fluidity under the action of external heating, and mechanically shear the melt, followed by the plunger or screw quickly entering the mould cavity with a lower temperature, and is cooled and solidified to form a plastic product consistent with the specifications [9]. The injection machine usually consists of an injection system, a mould clamping mechanism, a hydraulic system, an electrical control system, a heating part and other auxiliary parts such as a cooling part and a feeding part. The following mainly introduces the injection system, mould clamping mechanism, hydraulic and electrical control systems.

As one of the most essential aspects, the injection system is mainly composed of three parts: the plasticizing part, the injection part, and the pressure driving part. Its functions are as follows: (1) Plasticizing—the material is melted and plasticized uniformly by the combined action of the screw and the heating ring; (2) injection—the plasticized material is injected into the mould cavity at a set pressure and speed by the screw; (3) pressure is maintained—one molten material is injected into the cavity.

The screw remains stationary inside the cavity to replenish a portion of the molten material into the cavity, eliminating shrinkage caused by cooling, ensuring product quality, and preventing material from flowing back. The injection part consists primarily of a pressurising and driving device. When the required amount of melt is reached in the metering section, the screw stops rotating and remains stationary. The molten material in the metering chamber is then injected into the closed mould cavity via the injection cylinder, and the injection process is completed. The pressure device primarily supplies power to the screw, causing it to exert pressure on the material. Hydraulic pressure and mechanical force are the two main sources of power. At the moment, the majority of them rely on hydraulic pressure and a self-contained hydraulic system to supply pressure. The driving device primarily provides power to rotate the screw in order to complete the material's plasticization. AC motors and hydraulic motors are the most common drive devices. Hydraulic motors are more prevalent. The benefits of using a hydraulic motor drive are (1) smoother transmission characteristics, smaller starting inertia, and protection of screw overload; (2) smooth screw speed adjustment; (3) small size, simple structure, and easy service. The mould clamping device, which is one of the main components of the injection moulding machine, is made up of a mould closing device, a mould adjusting device, and an ejection device. The function of the clamping system is as follows:

(1) to support the forming mould and ensure that it has an ideal speed change process when reciprocating between the open and closed positions, to avoid strong vibration during operation, to prevent the mould from being damaged by impact, and to achieve smooth ejection of products and increase productivity.
(2) Provide enough clamping force to ensure that the mould can still be reliably locked under the action of the melt pressure to prevent cracks in the slit and thus ensure product quality. To that end, the clamping parts and moulds must be both strong and rigid.
(3) The product is cooled, the mould is opened, and a removal method is provided. The clamping device consists primarily of a fixed template, a draw rod, a power source, and other components. The hydraulic system's working quality not only determines the injection moulding machine's technical performance, but it also has a direct impact on the product's quality and the injection moulding machine's energy consumption.

The hydraulic system's composition varies depending on the operating cycle and performance requirements of the injection moulding machine, but it generally consists of a power source (hydraulic pump), control elements (pressure valve, directional valve, flow valve, etc.), actuators (hydraulic cylinder, hydraulic motor, etc.), and auxiliary devices (oil tank, oil gauge, etc.). The hydraulic circuit controls the actuator to complete the adjustment of each action programme and parameter of the injection moulding machine.

The injection moulding machine is distinguished by the following features: the moulded product has a complex shape. It can form metal insert products with good assembly and interchangeability; convenient and efficient product mould replacement, which contributes to increased product competitiveness; automation, simple operation, and high efficiency. Modern injection moulding machines are improving in terms of quality, efficiency, and consumption. Small and medium-sized injection moulding machines are geared toward precision moulding, whereas large-scale injection moulding machines are geared toward low energy consumption, quiet operation, stability, and efficiency. Engel created the duo 500 pico series injection moulding machine in 2015 (Figure 7.1). This injection moulding machine series employs a multi-variable pump system and some relatively advanced control technologies, resulting in a very short injection moulding time. This reduces the empty cycle time to 2.6 seconds, significantly increasing production efficiency, and is currently the most advanced injection moulding machine in the world. Furthermore, because the

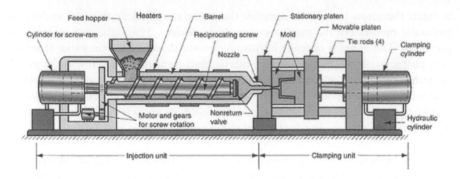

FIGURE 7.1   Injection Moulding Process [3].

duo 500 pico series employs multi-variable pump technology, energy loss is minimal. Battenfeld introduced the HM400/3400 Advantage model with a clamping force of up to 400t in 2016. The high-efficiency and energy-saving double pump system is used in the design process of the HM400/3400 Advantage model. When compared to other similar types of equipment, this device's maximum injection speed is obvious, reaching up to 300 mm/s. The Arburg S series injection moulding machine is designed with dual-pump and multi-pump control technology to meet the needs of simultaneous operations in order to improve working efficiency and meet a wide range of work requirements.

7.3.   DIFFERENT TYPE INJECTION MOULDING PROCESS

7.3.1.   Precision Injection Moulding

According to processing accuracy, injection moulding technology is divided into precision injection moulding and conventional injection moulding. In terms of repeatability, parameter control, and size control, precision injection moulding outperforms conventional injection moulding. Michaeli et al. [10], for example, conducted extensive research on the factors influencing the precision injection moulding process of aspherical plastic lenses, such as mould design, process parameters, and plastic raw materials. They performed actual injection experiments using various injection process parameters and mould structures, and analysed the degree of influence of process parameters and moulds on the imaging quality of aspherical plastic lenses. The final experimental results show that, when compared to mould design, the injection moulding process parameters have a greater impact on the imaging quality of aspherical plastic lenses. Zhang et al. [11] used injection moulding simulation software to optimise the design of moulds for large injection moulding products based on process monitoring and a series of experiments and verification, successfully predicting the deficiencies of microscopic features. The typical microfluidic chip was successfully replicated using injection moulding technology to obtain suitable process data and simulation parameters. This method greatly improved the product's accuracy. Figure 7.2 depicts the process using simulation software to predict and validate through experiments.

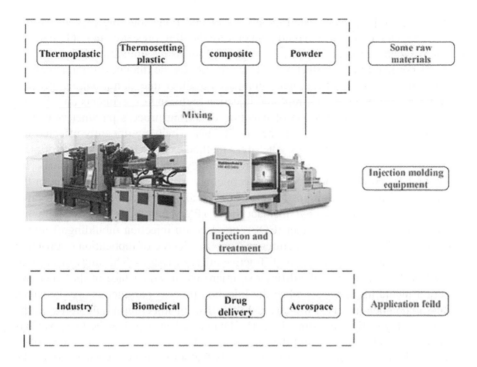

FIGURE 7.2 Advanced Injection Machine [11].

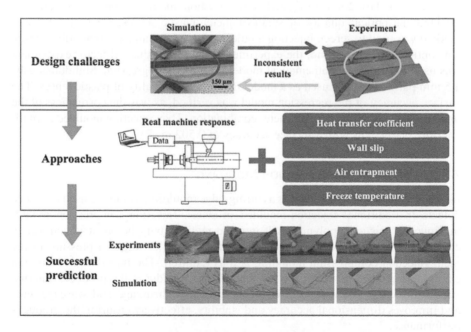

FIGURE 7.3 Process Monitoring and Experiment Prediction [11].

Young et al. [12] investigated the effects of injection moulding process parameters like melt temperature, mould temperature, holding pressure, and holding time on the residual stress and warpage of plastic lenses using simulation software. The study discovered that the effect of holding pressure on the thickness direction of the plastic lens is more significant when compared to other injection moulding process parameters. Macas et al. [13] repeated injection moulding experiments and discovered that an incorrect selection of injection moulding process parameters caused residual stress to be generated in the plastic lens, which significantly affected the structural size of the plastic lens. The stress method extends the service life of plastic lenses and improves their dimensional accuracy. Holthusen and colleagues [14] used single-point diamond processing technology to achieve high precision in the processing of diffractive microstructures in injection mould cores. They used three different plastics (poly(methyl methacrylate)) (PMMA), cyclic olefin copolymer (COC), and cyclo olefin polymer (COP) for precision injection moulding of mould cores with diffractive microstructures to study the degree of replication of different plastic materials on the surface of diffractive microstructures. The study discovered that PMMA is superior to the diffractive microstructured surface of the other two injection moulding materials when the injection process parameters are the same. Aidibe and colleagues [15] investigated the effect of injection mould cores on the surface quality of light-emitting diode (LED) plastic lenses processed using various methods. The study discovered that different surface textures cannot be described by a single parameter. Roughness isn't the only factor to consider when assessing the surface quality of plastic lenses. Simultaneously, an important parameter for accurately representing the quality of the plastic lens surface under different processing techniques was proposed.

Tasi et al. [16–20] investigated how injection moulding process parameters affected the optical quality of small-diameter plastic lenses. First, the Taguchi method was used to screen injection moulding process parameters that affect plastic lens optical quality. Second, the experimental data from the full factor experiment was used to develop a mathematical model for multiple regression. Simultaneously, a 5 mm plastic lens was used as a model to predict the quality of plastic lenses. The surface accuracy of the regression model was verified, as was the correctness of the injection moulding process parameter combination. They injection moulded a small-aperture plastic lens with a surface accuracy of 0.502 m (PV).

7.3.2. MULTI-MATERIAL INJECTION MOULDING

Injection compression moulding is a cutting-edge moulding technique used to reduce defects in traditional injection-moulded products. Injection compression moulding involves injecting polymer melt into the mould cavity before it is completely closed, and then applying uniform compression force or uniform pressure to the melt. Injection compression moulding is thus required. The injection pressure and clamping force are relatively low, which effectively reduces the product's internal stress, molecular orientation, birefringence, uneven shrinkage and warpage, etc., and improves dimensional accuracy and stability, effectively ensuring the product's performance.

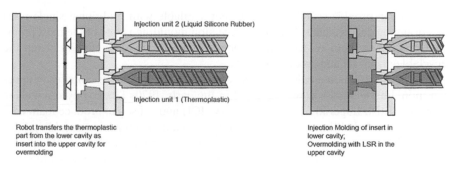

Injection unit 2 (Liquid Silicone Rubber)

Injection unit 1 (Thermoplastic)

Robot transfers the thermoplastic
part from the lower cavity as
insert into the upper cavity for
overmolding

Injection Molding of insert in
lower cavity;
Overmolding with LSR in the
upper cavity

FIGURE 7.4   Multi-material Injection Moulding Process [27].

When the cavity is closed, injection compression moulding leaves a certain gap, making the thickness of the cavity larger than the thickness of the product, and then fills the cavity with the melt. Because the cavity is wide at the start and the internal pressure is low, lower injection pressures can be used at this time. Following the completion of the injection, the injection moulding machine clamping system performs secondary clamping in accordance with the control requirements. To move the movable template forward, the clamping force is increased. Finally, the desired shape is formed, and cooling begins. When the product has finished cooling, it can be opened to remove it [21]. Injection compression moulding, which is more suitable for moulding precision optical devices, can reduce warping deformation and refractive index changes while improving product dimensional accuracy. It is gradually replacing injection moulding as the preferred method for producing plastic optical parts. During the compression phase of injection compression moulding, Barnes et al. [22] developed and experimentally validated a non-linear viscoelastic theoretical model. Kim and Isayev [23] investigated the differences between conventional injection moulding and injection compression moulding using the Leonov model.

According to research, injection compression moulding can effectively reduce the birefringence of the part when compared to conventional injection moulding. Kwak et al. [24] used neural network technology in computer simulation to improve the surface dimensional accuracy of optical components during injection moulding processing, and then performed experiments to validate the computer simulation results. Loaldi et al. [25] used a method to check the Fresnel surface microstructure and process, as well as a detailed evaluation of uncertain factors, to effectively establish the correlation between microstructures and improve the overall optical performance of the Fresnel lens. Bickerton and Abdullah [26] investigated the injection stage of injection compression moulding and discovered that it can significantly reduce injection time while increasing clamping force and injection pressure.

7.3.3. Second Injection Moulding

Two-shot injection moulding involves injecting two different (but chemically and rheologically compatible) polymers into the mould cavity at the same time or in a specific order [28, 29]. Products made from the combination of two different materials can provide special properties not available with traditional single resin injection moulded products. Secondary injection moulding is one of the most promising processing methods from an economic and ecological standpoint, given the recyclable materials and production costs.

Secondary injection moulding technology, also known as multi-component injection moulding technology, was invented in the early 1970s [30, 31] by researchers from the British Imperial Chemical Industry (ICI) and is used to produce foam core layers and solid shell layers of shaped products. Sandwich injection moulding (also known as core layer injection moulding), overlap injection (overlay) moulding, and two-color injection moulding are all examples of secondary injection moulding.

Secondary injection moulding technology is classified into four types based on their processes: single-runner injection moulding, dual-runner injection moulding, triple-runner injection moulding, and mono co-injection moulding. It is classified as sequential co-injection moulding or sandwich injection moulding based on the injection sequence of each component in the moulding process. It is the most basic and second-oldest multi-component injection moulding process in industrial production [32, 33]. Sandwich injection moulding is primarily used to control different types of materials in a specific order. The shell of the product is moulded from the first injection of the material through an injection moulding nozzle, and the core layer is moulded from the second injection of the material. Sandwich injection moulding is an important moulding technology. One injection moulding process can form products with a shell layer and various types of core plastics. The following are the reasons for the widespread acceptance of this moulding process: (1) it can be used to create shells and core layers with varying hardness, low shrinkage, and good gas barrier; and (2) the technology can be applied in new fields such as waste recycling. Core injection moulding can produce both large thick-walled parts and traditional thin-walled parts depending on the core material chosen, such as low-density plastic with a structure similar to foam materials. In addition, the application of these low-density plastics can eliminate many defects caused by moulding, such as shrinkage, deformation, cracking, and internal stress [34].

The injection process's core injection moulding is detailed here. Material A is injected into the cavity's surface to form a shell layer, and then material B is injected into A to form the product's core structure. To achieve a stable product with a good cross-section, the shell and core materials should have similar rheological properties, such as a similar melt index and moulding conditions. If two plastics are injected, low-viscosity molten plastic will wrap around high-viscosity molten plastic. Furthermore, the shell layer thickness is determined by the product wall thickness, melt viscosity, and moulding conditions. The primary goal of using this moulding process is to create low-cost, high-performance products by combining two plastics. The chemical affinity and shrinkage behaviour of these two materials must be similar in order to achieve this in a core mould. Zang et al. [35] used Mouldflow to investigate the effect of core injection moulding material viscosity and process parameters on the penetration depth and distribution uniformity of the core layer. The findings demonstrate that as the core layer melt's homogeneity decreases and the viscosity ratio between the core layer melt and the shell layer rises, the core layer melt's penetration length also decreases. Research also found that the core layer injection speed, melt temperature, and mould temperature have a greater effect on the penetration depth and distribution uniformity of the core layer. Overlap injection moulding is plasticizing two different plastics in two injection moulding machine barrels, and then sequentially injecting them into the mould cavity [36]. The first injection part is called the matrix material, and the latter injection part is usually semi-coated or coated. On the base material, plastic products of different colors or textures are formed, which is also called the "in-mould assembly" or "in-mould welding" moulding process.

Second injection moulding has a significant cost and quality advantage over traditional injection moulding [37–40]. Secondary injection moulding is better suited for the production of thick-walled products than traditional structural foam because it has a higher apparent quality. Combining different materials can achieve combination properties that are difficult to achieve with a single resin. Furthermore, the use of low-cost materials throughout the core structure can reduce product costs. Two-injection moulding is one of the most effective production technologies for achieving high recycled material utilisation, repeatability, and high-quality products.

7.4.   INJECTION MOULDING PROCESS PARAMETERS

In the injection moulding process, high-molecular polymers are subjected to extremely complex heating–cooling and mechanical shearing. The crystalline morphology in the thickness direction of the part is also different due to the different thermal-cold history and mechanical shearing history of polymer regions, and it will show a clear skin-core structure with an obvious anisotropy. Different injection moulding process parameters produce different polymer microstructures. Polymer microstructure differences are primarily manifested in differences in molecular chain arrangement, molecular chain orientation, and crystallisation. The differences in these polymer microstructures will be different. Crystallization is also a significant factor in the evolution of the aggregate structure and microstructure of the moulded product, which has a significant impact on the product's physical properties and dimensional stability. Because of the incomplete crystallisation of the melt, the crystalline polymer contains two components: crystalline and amorphous regions [41]. The crystallinity of the polymer is a physical quantity used to analyze the quantitative data of the crystalline and amorphous regions. The mass percentage (or volume percentage) of the crystalline phase in a polymer containing both crystalline and amorphous regions is defined as crystallinity. The influence of process conditions on part performance is multifaceted. Parts moulded under different process conditions will have different internal molecular structures at the micron level, which may result in a variety of defects such as bubbles and internal stress shrinkage warpage at the macron level. A large number of studies have shown that injection pressure and melt temperature have a greater impact on the mechanical properties of the part, and that proper process control can effectively improve the mechanical properties of the product [42].

7.4.1.   MELT AND MOULD TEMPERATURE

By increasing the temperature of the plastic melt, the internal molecular chains will move faster, resulting in better relaxation. After cooling the product, the interaction energy between molecular chains will be reduced, the internal stress will be better released, and the internal stress will be smaller. Because of the low degree of shrink-age and deformation, the surface of the injection moulded part is smoother and more shiny. Wang et al. [43], for example, investigated the residual stress of thin-walled parts and discovered that mould temperature had a greater effect on the residual stress of the surface layer and core region, while pressure had a smaller effect. Wimberger et al. [44, 45] found that birefringence is related to the residual stress of the transpar-ent material itself and the product. Since the melt flow during the filling stage and the shrinkage abnormality is caused by the temperature and pressure difference during the injection moulding process, the molecular orientation will cause the component to generate internal stress. For example, Yao et al. [46] used the injection moulding simulation method to analyze the melt flow during injection moulding, and analyzed the effects of various process parameters and mould sizes on the injection moulding process. Under the condition that the mould temperature is low and the size ratio of the part is the same, for the par with a smaller thickness, it will normally require a greater maximum pressure and a longer injection time during injection. When the mould temperature is high and the size ratio of the part is the same, for the part with a smaller thickness, the maximum pressure required for injection does not change significantly, but a longer injection time is required. The maximum pressure required for injection moulding also does not change significantly, and the maximum pressure is smaller when the injection time is short.

7.4.2. INJECTION PRESSURE AND SPEED

The melt experiences strong shear action in the barrel during the injection moulding process, and the injection pressure and injection rate are two process parameters that determine the shear strength of the melt. Injection pressure and injection rate are two process parameters that are related. Within a certain range, increasing the injection pressure also increases the injection rate, resulting in increased melt flow rate and shear effect. Yang et al. [47] evaluated the refractive index change of injec-tion-moulded polymethyl methacrylate optical lenses at different filling pressures using a measurement system based on Shack-Hartmann wave front metrology. It was found that under different holding pressures, the refractive index changes of the optical polymer injection moulded lenses are larger, and the refractive index changes of the products under high holding pressure and lower holding pressure are more uniform. Yu et al. [48] studied the effect of process parameters on the properties of injection moulded parts using a variable mould temperature injection moulding. The results show that the injection speed and mould temperature have a greater impact on the surface replication performance of the part, indicating that the injection speed and mould temperature are the main factors affecting the surface gloss of the part. Chivatanasoontorn et al. [49] studied the effect of injection speed (i.e. shear rate) on the microstructure of the polypropylene (PP) surface layer. High injection speed (1000 mm/s) improves crystal orientation and the relative amount of phase crystals in the surface layer, according to the results (0–30 m). Micro-cutting and progressive loading scratch tests were used to investigate the effects of these microstructures on surface properties. The injection speed or shear rate increased the surface strength and scratch resistance, indicating the surface mechanical properties. This is related to highly oriented microstructures on the surface.

7.4.3. HOLDING PRESSURE AND TIME

The holding pressure stage's primary function is to perform shrinkage after the moulding is complete in order to avoid defects such as shrinkage due to volume effects caused by temperature reduction. Increasing the holding pressure causes the melt to shrink and increases the density of the part. Part of the melt is still flowing during this process, and this part of the melt will also be oriented due to the flow. Furthermore, the free volume must be reduced due to the increase in density, and the mutual restraint effect of the molecular chains in the melt is stronger. The molecular chains require more time to relax, and previously oriented molecular chains may reduce de-orientation. When the holding time is extended, the melt density rises and the movement of the molecular segments becomes more difficult. The size of the generated spherulites will be reduced, as will the haze, and the transparency will be increased. However, as the melt cools and the gate cools and freezes, the effective dwell time reaches a maximum and the dwell time continues to increase. The actual effective maximum dwell time will remain constant, as will the haze. Min et al. [50] studied the influence of the filling and compression processes on the residual birefringence structure of the optical disc by testing the birefringence distribution and the extinction angle. Two anomalous birefringence and extinction peaks near the center in the thickness direction indicate the holding pressure, which indicates that the holding pressure has a certain effect when the flow is replenished during the holding phase. In addition, the injection/compression process has a more uniform birefringence distribution than products obtained by conventional injection moulding. Weng et al. [51, 52] studied the residual stress and birefringence of micro-lens array precision injection moulding through simulation and injection moulding experiments. The consistency was demonstrated by the residual stress and birefringence distributions obtained from simulation and experiment. The maximum residual stress is always found near the gate, and the maximum value of the residual stress decreases as the holding pressure and cooling time increase. The holding time has little effect on the maximum value of the residual stress. The mould temperature is the most important process parameter that influences residual stress, and higher mould temperatures result in a lower maximum residual stress value near the gate. Lee et al. [53] used numerical simulation to predict the birefringence distribution of flow-induced and thermally induced residual stresses from the gate to the centre. Isayev et al. [54] used a combination of finite element method (FEM) and finite difference method (FDM) for physical modelling and two-dimensional numerical simulation of injection moulding. Residual flow birefringence in injection-moulded parts is simulated by using a compressible nonlinear viscoelastic constitutive equation while considering the filling, packing, and cooling phases of injection moulding. The simulation results show that in the absence of shrinkage, the birefringence of the surface layer of the part is mainly formed by flow stress cooling and solidification, and the core region is formed by thermal stress cooling and solidification. In the absence of shrinkage, as the shrinking melt continues to flow, a second birefringence peak forms between the product centre and the maximum birefringence position, and the simulation results agree with the experimental test results. Pantani et al. [55] investigated the effect of various holding pressures on the molecular morphology distribution, and discovered that holding pressure has a greater effect on molecular orientation. Furthermore, a molecular structure distinct from that observed under normal pressure was observed under high holding pressure. The molecular orientation under pressure decreases with increasing distance from the surface layer, and the skin-core structure differs. The molecular orientation of the high-pressure-preserving pressure component maintains a high value throughout the thickness, and even a second maximum value except the surface layer is observed.

7.4.4. APPLICATION OF PROCESS PARAMETERS

The mutual coordination of the injection moulding parameters of the injection moulded products can not only keep the total product quality unchanged, but also achieve a reduction in energy consumption and improve product economics. For example, by interacting with cooling time, screw speed, mould temperature, and nozzle temperature, the total mass of the part can be maintained and the cycle time of the product can be effectively improved, reducing energy consumption in the manufacturing process and improving the economics of the injection moulding process [56]. The mould must be heated during the injection moulding process. Using this feature, injection moulding can also optimise the temperature of the hot runner and nozzle by controlling the coolant flow of the mould, resulting in an effective reduction of energy consumption and improvement of the injection moulding process without sacrificing part quality [57]. To summarise, injection moulding is a multi-step process. In the actual production process, the injection moulding parameters can be adjusted to find the optimal parameters based on the properties of different materials. This not only maintains total injection quality, but also increases the economic benefit of the injection moulding process.

7.4.5. INJECTION MOULD AND INJECTION PROCESS PARAMETERS

Many scholars have done a lot of research on injection mould and injection process parameters. For example, Yang et al. [58] analyzed that in the process of mould forming, when the plastic parts have an arc structure, the use of ring gates for injection moulding design will have a relatively large impact on the process parameters and precision of the plastic parts. The results show that the flow pressure holding stage has a great influence on the straightness and roundness of the plastic parts. Chang et al. [59] studied the filling process of injection moulding by injecting the resin reinforcement layer into the cavity area of the mould because there were certain gaps between the moving mould and the fixed mould during the closing stage of material moulding. The results showed that the filling time could be greatly reduced by speeding up the compression rate and narrowing the gap size. Lee et al. [60] conducted a thorough analysis of the filling and flow pressure retention process, ensuring not only that the pressure distribution of the mould cavity was even, but also that the melt filled each mould cavity at the same time, ensuring that the casting system achieved the best and achieved a balance in many aspects. To improve replication efficiency and reduce costs, micro-injection moulds (WM) and micro-injection compression moulds (ICM) are commonly used to replicate the surface microstructure of thermoplastic polymer. The microstructure is the primary manufacturing component derived from traditional injection moulds. Christiansen et al. created a sawtooth microstructural surface with a specific angle and successfully realised the microstructural region's anti-light reflection function [61]. Optimization of cell IM injection moulding process parameters can significantly improve the replication quality of microstructures, thereby improving cell adhesion and proliferation in the polymer substrate [62]. A rapid thermal cycling system has been successfully developed based on the characteristics of micro-injection moulding, which can quickly control the mould temperature to improve the replication quality of microstructures [36]. The vacuum mould exhaust method [63] is another helpful technique that aims to reduce the air resistance of the polymer into the microstructure. Sorgato et al. discovered through systematic research that vacuum technology can contribute to the improvement of microstructure replication quality only when the mould temperature reaches a certain value; otherwise, it has the opposite effect [64].

7.4.6. Composite Material Injection Moulding

Composite materials are widely used in industrial production and have gradually become one of the most important indicators of a country's scientific, technological, and economic strength. Advanced composite materials not only have excellent properties such as high strength and heat resistance, but they are also widely used in many fields such as aerospace, transportation, and machinery. Composite injection moulding is a current process method for moulding composite materials. The current injection moulding method is the primary method for moulding composite materials. It is primarily made by injecting metal elements such as Ti, Mg, Al, and other metal materials, as well as matrix resin composite materials [65]. The injection moulding process for most composite materials consists primarily of the four working steps listed below. The first step is to dry the selected composite material at a specific temperature and in a specific environment for two to three hours. The dried composite material is then placed in a suitable injection moulding machine, which is then heated to a suitable temperature to make the material appear molten. Plasticizing temperatures range from 230 to 250 °C, and the plasticizing cycle lasts 15 to 20 seconds. The third step is to use the injection screw to inject the molten composite material into the mould and then cool the composite material in the mould to make the composite material with the mould cavity fixed to obtain the corresponding parts. It should be noted that when the composite material is injected for 1–2 minutes, the injection should be performed after a 1–2 minute interval. The mould should be cooled for 40–50 seconds after injection. The moulded product must be removed in the fourth step. In the actual manufacturing process, an ion air gun can be used for rapid sweeping, and a composite material product is obtained [66].

Mitsubishi Heavy Industries and Nagoya Machinery Works pioneered composite material injection compression moulding (IPM) technology. Overall compression and partial compression are the two broad categories of composite injection compression moulding technology. Overall compression injection moulding technology refers to the process of filling a resin into a model while keeping the model open to a certain degree [67]. The cylinder is then compressed to move the mould until it is completely closed. The resin is then refilled. The entire compression process is dependent on the product moulding surface on the stencil rather than the entire moving film plate. The main benefit of this process is that it can produce thin and light products at low pressure [68]. Composite injection compression moulding technology is generally appropriate for relatively poor fluidity, but thin-walled products on the outside, such as polymer compounds PC and fibre-filled engineering supplies. Furthermore, the company's composite material direct injection moulding technology is primarily applicable to materials containing higher concentrations of glass fibre, C fibre, organic powders, or inorganic powders [69]. This method, for example, can be used to perform injection moulding of commonly used calcium carbonate composite materials as well as injection moulding of wood flour composite materials [70]. To highly disperse the fillers on the matrix resin in the preparation of composite materials, the traditional method requires the matrix resin and the glass fibre to be fully mixed, and the assistance of a twin-screw extruder is required to perform out-of-process mixing. The moulding method produces pellets and then the corresponding products. This method uses a lot of energy and results in extensive resin degradation, oxidative discoloration, and excessive shearing of the corresponding fibres. Direct injection moulding does not necessitate pelletization with an extruder. A product can be obtained by blending it directly into the mixture, because injection moulding machines are mostly single-spiral devices with small diameters. Therefore, for injection moulding technology, the most important thing is to improve the working efficiency of the screw rod [71, 72]. The injection moulding technology of composite materials can be divided into resin-based composite material injection moulding, metal-based composite material injection moulding, and cement-based composite material injection moulding according to the different matrix materials used. They are introduced one by one in the following paragraph below.

The five steps in the moulding process of resin-based composite products are as follows. The first is the hand lay-up moulding process, which uses manual or mechanical assistance to cover the resin-based composite material on the mould. This operation is extremely simple, but it takes time and is not suitable for mass production [73, 74].

The second is an injection moulding process, which uses a spray gun to spray resin-based composite materials onto the corresponding mould and mould the product on the mould. This moulding method is suitable for both short fibres and resin injection. The injection moulding process has a low cost and small losses, but uneven curing and pollution are common [75]. The compression moulding process is the third. This process produces a more beautiful, efficient finished product with very close size specifications. However, its procedure is relatively complex, and it is only appropriate for processing small and medium-sized products. The fourth process is winding moulding, which has a high production efficiency as well as a low cost. A concave-surfaced product, however, cannot be wound. The autoclave forming process is the fifth type. This process produces a better product, but at a higher cost. In order to use resin-based composite materials more efficiently, the automation and specialization of the process should be actively promoted, so as to improve the development of injection moulding technology [76]. With the emergence of the energy crisis, light weighting is one of the main development directions of industrial products. The body of the car, the floor and the brake parts, the aircraft frame and some avionics can be prepared by the resin-based composite material injection moulding.

7.4.7. Metal Matrix Composite Injection Moulding

Metal-based composite materials are composed primarily of metals or metal alloys as the main matrix, with metal or nonmetal materials used to create heterogeneous mixtures. According to the type of matrix, metal-based composite materials are classified into four types: aluminum-based composites, nickel-based composites, magnesium-based composites, and titanium-based composites [77]. Metal matrix composites' main performance characteristics are closely related to the properties of their reinforcements. They have good electrical and thermal conductivity in general, and they perform well at high temperatures due to their low thermal expansion coefficient. It is not easily absorbed by moisture, does not age easily, and has good air tightness in the processing operation. Due to their significant advantages, metal-based composites have become one of the topics of in-depth research by materials researchers. Simultaneously, there are some issues at the interface of metal-based composite materials. As a result, in the field of interface research, more emphasis should be placed on the development of advantageous analysis methods. Existing interface reaction methods are optimised, and more effective measures are proposed to truly promote metal matrix composite development [78]. Because of the variety of metal-based composite materials, the main matrix and reinforcements will be very different, resulting in a variety of moulding processes. The following three methods are currently used in the specific moulding process of metal matrix composite materials. The first moulding method is a solid state method in which the metal matrix and reinforcement are solid and the entire process is carried out in a low-temperature environment. The second method is the liquid metal method. The metal matrix is molten at this point, and a composite moulded product with solid reinforcement is formed. Because the preparation requires a higher temperature, the interface reaction must be strictly controlled, which is also a key factor for the success of this method. The third method is the self-generating method and other manufacturing methods, the main principle of which is that in the metal, an appropriate amount of a reaction element is added to generate a solid reinforcing phase through internal reaction. Furthermore, elements can be deposited on the composite material layer through electroless plating or electroplating [79, 80].

7.4.8. Cement-Based Composite Material

Fibre-modified cement-based composites are the most worthy of popularisation in the application process because they can improve tensile properties and flexural strength [81]. Furthermore, it can reduce shrinkage and cross-section size, making the component lighter. Because of their different modes of action, fibre-modified cement-based composites are classified according to different fibres [82]. As a result, it's classified into four types: short fibres, network fibres, profiled fibres, and surface-modified fibres. Fibre toughness can be increased by changing the type of fibre and improving its properties and adhesion. Fibre-reinforced cement materials are classified into four types.

The first is a steel fibre cement-based composite material that was used for the first time in 1960. It has good impact resistance and sufficient toughness as its main performance characteristics. In specific application practice, steel fibre has a better effect on cement-based composite materials than other types, and its performance has an advantage. It has been widely used in the construction of airport runways, bridge deck paving, ocean engineering, ballistic engineering, and other projects. The second type of composite material is glass fibre-reinforced cement-based composite material, which is ideal for wrapping steel structures to improve fire resistance. Because the material prevents cracking, it also prevents rusting. As a result, it is ideal for applications in the field of marine structures. In general, glass fibre is made by reacting quartz sand, dolomite, paraffin, and other materials with strong acids and alkalis. In some cases, oxides are added to aid in the preparation, such as aluminium oxide, titanium dioxide, and the like [83]. This type of material is the most commonly used fibre in the specific application process, and it is also an essential component of high-tech research and development. Its elasticity and stretchability are also relatively good due to the ease with which its raw materials can be obtained. As a result, it is also known as one of the high-performance reinforcing materials. Glass-fibre-reinforced cement-based composites can be classified in a variety of ways. Among them, according to the alkali content in the raw materials, it can be divided into four categories: alkali, medium alkali, low alkali, and alkali-free glass fibre reinforced cement-based composites.

Polypropylene (PP) fibre-reinforced cement-based composites are the third option. The material's appearance improves permeability and both wear resistance and impact resistance to the outside world. Aerated cement concrete and sprayed cement concrete are two applications. The fourth type of composite material is smart cement-based composite materials, which refer to the composite smart components found in traditional cement-based materials. Sensors or actuators, for example, are added to certain cement-based materials. Navigation cement-based composite materials, self-repairing cement-based composite materials, and temperature monitoring cement-based composite materials are examples of intelligent cement-based composite materials that have been used recently. One of them is navigation. Cement-based composites are commonly used in automobiles and can be used to determine the specific driving route of automobiles using electromagnetic waves [84, 85], whereas self-repairing cement-based composites incorporate hollow capsules containing binders in composites. Cracking can be repaired. The electrothermal and thermoelectric effects are primarily used in self-monitoring cement-based composite materials, and monitoring is carried out in response to changes in internal temperature. Automatic snow and ice removal on airport runways and high-speed roads are two examples of specific applications [86].

7.4.9. Carbon Composite Injection Moulding

Carbon composite is a low thermal expansion index material made from carbon fibre and resin. It also has a high strength, resistance to fatigue, and other properties. Hand lay-up, injection moulding, winding, and resin transfer moulding are the four methods for processing carbon composite materials. The winding method is the most common and appropriate method for producing pipes. Carbon composite materials can be injection moulded using either the impregnation method or the CVD method. In general, there is still room for advancement in the field of composite materials science [87]. Carbon composite materials have the inherent properties of carbon materials as well as the softness of textile fibres. Because the carbon valence is relatively stable, it has excellent acid and alkali resistance properties. Among its physical properties, it has a low friction coefficient, which contributes to the lubricity of carbon composites. Carbon fibre composite materials can be used for bridge deck reinforcement, tunnels, and plant engineering maintenance in specific applications. The basic principle of carbon fibre composite material injection moulding is to precisely measure the two reactants, mix and collide them under high-pressure conditions, and then fill the mould. Inside the mould, the mixture is thoroughly mixed, and a rapid polymerization reaction occurs to produce the cured product [69]. It should be noted, however, that during this process, reaction monomers and reinforcing materials must be mixed together in the mould to prevent the quick reaction. According to specific research, carbon fibre is a high-tech material whose products outperform those of many other materials. At the moment, most carbon fibre composite material research is focused on preparation technology and component analysis [88]. With the market's continued development, the use of carbon fibre composite materials will become more common. Carbon fibre composite material development has a significant impact on moulding technology and even the carbon fibre industry.

7.4.10. Other Composite Injection Moulding Technologies

The changes in the matrix must be considered when developing composites. The soft matrix is created first, and then the hard matrix is gradually transformed. A new type of ceramic matrix is now available, having evolved from resin to metal. The ceramic matrix is commonly referred to as polyphase conformable ceramic, and the shift in ceramic material research from monolithic to multiphase will open up more options for ceramic material design. By incorporating one or more other components into the base material to form ceramic matrix composites, significant improvements in mechanical properties of single-phase ceramic materials have been achieved (CMCs). TiC, TiN, TiB2, SiC particulate, SiC whisker, B4C, ZrO2, WC, Ti(C,N), Cr3C2, NbC, and other particles or whiskers are commonly used as reinforcing components. Ceramic composites are gaining popularity, with oxide matrices, particularly alumina, dominating. However, the corresponding material compositions, processing techniques, reinforcing and toughening mechanisms, properties, and applications require further investigation [67, 69, 88–90]. Its toughness will be increased because it is a second-phase material introduced into the ceramic matrix [91]. Ceramic matrix composites have low thermal conductivity and high wear resistance, high temperature resistance, and corrosion resistance [69]. Furthermore, it has evolved into the most ideal high-temperature structural material today. Ceramic matrix composites, according to relevant research, will gradually become the material of choice for hot-end structures. As a result, many countries have established related research in this field in the hope of promoting composite material development.

Of course, there are many different types of composite injection moulding, such as self-reinforcing single polymer composites created by co-injecting the same types of polymers with different properties into an object [68]. Traditional injection moulding has a higher specific strength, no interface heterogeneity, and can be recycled easily. Foam injection moulding (FIM) can be used to create foamed PP/polytetrafluoroethylene (PTFE) [92], fibril PLA/PET composites [93], PLA with PTFE nanofibrils and polyhydroxyalkanoates (PHA). Other injection composite material components produced by special injection moulding processes are also examples of injection composite materials produced by reinforcing each other [94]. Of course, other fillers are added to the base material to create new composite materials via special injection moulding methods, such as the manufacturing of isolated CNTs/high-density polyethylene (HDPE)/ultra-high molecular weight polyethylene (UHMWPE), which is used to prepare conductive composite materials [94] and PP-glass fibre/carbon fibre hybrid injection moulded composite materials prepared via direct fibre feed injection moulding (DFFIM).

7.5. MICRON INJECTION MOULDING

In the 1980s, the concept of microinjection moulding was first proposed. The process of pushing molten plastic from an injection machine into a micron mould and cooling in the mould to obtain a product under the action of a plunger or a screw is referred to as micron injection moulding. After the material has been plasticized in the injection moulding machine's heating barrel, the plunger or reciprocating screw is injected into the cavity of the closed mould to form the product. Injection moulding is characterised by the ability to process complicated shapes, precise dimensions or products with inserts, and high production efficiency [95]. Microinjection moulding products are used in optical communications, computer data storage, medical technology, biotechnology, sensors and actuators, micro-optical devices, electronics, and consumer products, as well as equipment manufacturing and mechanical engineering, thanks to the rapid development of injection moulding precision technology. The number of applications in this field is growing. Injection moulded products are commonly used in the following applications: watch and camera components, car collision, acceleration, distance sensors, hard disc and optical drive read/write heads, micro-pumps, small spools, high-precision gears, pulleys and coils, optical fibre switches and connectors, micro-motors, and so on [96, 97]. The micro injection moulding machine has three independent mechanisms: a plasticizing screw, a metering piston with an embedded pressure sensor, and an injection plunger. Because the injection unit is driven by a cam mechanism, it has excellent injection and stopping response. As a result, this machine is suitable for producing micro parts with high accuracy and stable moulding [5, 97].

IM is commonly used for cosmetic/pharmaceutical packaging and, more recently, for the production of biomedical devices such as scaffolds and microneedles. In the development of portable micropump delivery systems for chemotherapeutic drugs, insulin, or immunisation agents, promising results were obtained in the manufacturing of microfluidic devices. Speiser was the first to propose using IM to prepare drug dosage forms. The main factors influencing the success of this technique in the pharmaceutical industry are its scalability and patentability. Indeed, IM is a potentially automated cyclic process (continuous production) that can be easily scaled up to the industrial scale by using larger equipment and moulds. A single IM cycle can last only a few seconds, and in many cases, moulds allow for the concurrent production of multiple units, reducing process time even further. Because of the versatility of the IM technique, drug delivery systems with specific shape and/or dimension properties can be created. Furthermore, no solvents are required, saving time and money while maintaining stability. Moreover, the typical process conditions, pressure and heat, both reduce microbial contamination (auto sterilisation) and promote drug–polymer interactions, potentially resulting in solid solutions or dispersions. This, like the Hot Melt Extrusion (HME) technique, would increase dissolution rate and, possibly, improve bioavailability of poorly soluble drugs [19, 20]. The multiplicity of patents filed over the last ten years demonstrates the great potential of IM for producing drug delivery systems, even though the number of products in advanced development, or which are already on the market, is still limited (e.g. Capill®, ChronocapTM, Egalet®, SeptacinTM); thus, there is still room for improvement and in-depth investigation.

7.6.   CONS OF INJECTION MOULDING

In terms of quality, injection moulding beats out 3D printing as it is currently used. There are two reasons for this. For starters, injection moulded parts are less prone to delamination than 3D printed parts. This is due to differences in manufacturing details.

Injection-moulded parts are made of a single piece of homogeneous material that cures as a single piece. There is no layering. This stands in contrast to FDM materials, which are built in layers. Furthermore, 3D printing is primarily used for prototypes rather than finished products.

Because 3D products are assembled in layers, the surface of 3D printed objects tends to be slightly rougher than that of injection moulded objects. However, an additional smoothing finish during 3D printing aids in the resolution of surface issues.

The better time option is determined by the intended volume of production. If you only need a few copies, 3D printing is the better option in terms of time. For small-scale production, the time spent building and operating the mould is hardly justifiable.

3D printers are extremely detailed. This enables precise creativity and adjustment. However, it has an impact on production time. As a result, 3D printers may require more time to produce an object than injection moulding. 3D printers, on the other hand, are easier to set up than injection moulding machines. This initial time-saver gives 3D printers an advantage in small production volumes.

Injection moulding saves time in large-scale production. This is especially true when the product is meant to be the final version.

Injection moulding involves melting the material and then delivering it into the mould cavity while it is still molten. It cools and hardens there. It involves fewer steps than 3D printing. At the end, it usually delivers the final version of the product.

In 3D printing, the material is melted but not completely molten. The material is then added layer by layer. In contrast, injection moulding does not require a layering process. The product delivered is usually not the final version; rather, it is a prototype.

3D printers make it easier to make production adjustments. In contrast, injection moulding has fixed design features of the moulds. As a result, 3D printers provide greater product delivery flexibility than injection moulding.

3D printing also allows for the creation of more complex designs. This is far more feasible than injection moulding.

Injection moulding is effective with a wider range of materials, including nylon, acrylic, polycarbonate, polyoxymethylene, polystyrene, metal, and others.

3D printing also works well with a wide variety of materials. Plastics, resins, and, more recently, metals are popular materials for 3D printing. Both modes of production have similar material options. However, if you want to manipulate materials creatively, 3D printing is a better option.

Some factors influence whether 3D printing or injection moulding is the more cost-effective option. The most important is the number of parts to be manufactured. If you only need a few copies, such as prototypes, 3D printing is the best option.

For small-scale operations, injection moulding is expensive. This is due to the cost of designing the mould, including the precise cavity. However, for high-volume production, such as 100+ parts, injection moulding becomes the more cost-effective option.

The cost of 3D printing is lower for entry-level projects. The cost of acquiring a working 3D printer is insignificant in comparison to the cost of an injection moulding setup. There are also open-source software communities and other support networks for 3D printing that can help reduce costs.

Injection moulding is prohibitively expensive to operate and maintain for production purposes. There's also the fact that it allows for a lot of trial and error without the high cost [98–101].

7.7.   BENEFITS OF 3D PRINTING

Using additive manufacturing provides numerous benefits, making it a viable alternative to conventional manufacturing techniques.

For starters, 3D printing is ideal for iteration. For a long time, additive manufacturing was thought to be a technique for rapid prototyping. It is no longer the case, as it has also become a very reliable manufacturing technique. However, one of the best advantages of this manufacturing technique is the ability to quickly and easily prototype. When using additive manufacturing, one can print the project to test it, then modify it using 3D modelling software and print it again to validate the changes. The adaptability of this process will not only save time and money, but it will also help improve the product design.

Wasted material is also decreased by using 3D printing. When compared to procedures like CNC machining, injection moulding was originally regarded to be a low-scrape production method, but 3D printing now seems to be the victor in this division! When using 3D printing, you only utilize as much material as you require for the complete project.

(N.B. You can also avoid inventory problems and costs by improving your storage and supply chain management with 3D printing.) You can, indeed, 3D print your 3D design whenever you need it! You don't have to worry about storage!

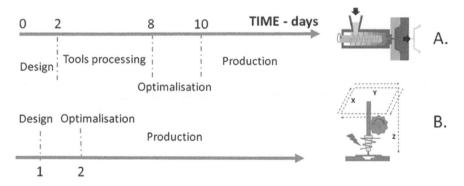

FIGURE 7.5 Time Duration of Both Injection Moulding and 3D Printing [103].

It avoids the crowded warehouses; the use of 3D printing for manufacturing may result in the dematerialization of the supply chain. It could be an opportunity to rethink the company's entire manufacturing process and reduce tooling investment. What exactly is digital inventory? This solution is as simple as 3D printing your parts on demand. There's no need to keep the part; simply print it when you need it [102].

REFERENCES

1. Zema, L., Loreti, G., Melocchi, A., Maroni, A., & Gazzaniga, A. (2012). Injection moulding and its application to drug delivery. *Journal of Controlled Release, 159*(3), 324–331. https://doi.org/10.1016/j.jconrel.2012.01.001
2. Ebnesajjad, S. (2015). Injection moulding. *Fluoroplastics*, 236–281. https://doi.org/10.1016/b978-1-4557-3197-8.00010-9
3. Xcentricmould. (2022, March 22). Injection moulding process | Xcentric Mould & Engineering. *Xcentric Mould*. www.xcentricmould.com/injection-moulding-process
4. Ebnesajjad, S. (2003). Injection moulding. *Melt Processible Fluoroplastics*, 151–193. https://doi.org/10.1016/b978-188420796-9.50010-2
5. Hamidi, M. F. F. A., Harun, W. S. W., Samykano, M., Ghani, S. A. C., Ghazalli, Z., & Ahmad, F. (2017). A review of biocompatible metal injection moulding process parameters for biomedical applications. *Materials Science and Engineering: C, 78*, 1263–1276. https://doi.org/10.1016/j.msec.2017.05.016.
6. Dirckx, M. E., & Hardt, D. E. (2011). Analysis and characterization of demoulding of hot embossed polymer microstructures. *Journal of Micromechanics and Microengineering, 21*(8), 085024. https://doi.org/10.1088/0960-1317/21/8/085024.
7. Ye, H., Liu, X. Y., & Hong, H. (2008). Fabrication of metal matrix composites by metal injection moulding—a review. *Journal of Materials Processing Technology, 200*(1–3), 12–24. https://doi.org/10.1016/j.jmatprotec.2007.10.066.
8 Fu, H., Xu, H., Liu, Y., Yang, Z., Kormakov, S., Wu, D., & Sun, J. (2020). Overview of injection molding technology for processing polymers and their composites. *ES Materials & Manufacturing, 8*, 3–23. https://doi.org/10.30919/esmm5f713
9. Peças, P., Ribeiro, I., Henriques, E., & Raposo, A. (2019). Additive manufacturing in injection moulds—life cycle engineering for technology selection. *Advanced Applications in Manufacturing Engineering, 2019*, 105–139. https://doi.org/10.1016/b978-0-08-102414-0.00004-5.

10. Michaeli, W., Hessner, S., & Klaiber, F. (2009). Analysis of different compression-moulding techniques regarding the quality of optical lenses. *Journal of Vacuum Science & Technology B: Microelectronics and Nanometer Structures Processing, Measurement, and Phenomena, 27*(3), 1442–1444. https://doi.org/10.1116/1.3079765.

11. Zhang, H., Fang, F., Gilchrist, M. D., & Zhang, N. (2019). Precision replication of micro features using micro injection moulding: Process simulation and validation. *Materials & Design, 177*, 107829. https://doi.org/10.1016/j.matdes.2019.107829.

12. Young, W. B. (2005). Effect of process parameters on injection compression moulding of pickup lens. *Applied Mathematical Modelling, 29*(10), 955–971. https://doi.org/10.1016/j.apm.2005.02.004.

13. Macías, C., Meza, O., & Pérez, E. (2015). Relaxation of residual stresses in plastic cover lenses with applications in the injection moulding process. *Engineering Failure Analysis, 57*, 490–498. https://doi.org/10.1016/j.engfailanal.2015.07.026.

14. Holthusen, A. K., Riemer, O., Schmütz, J., & Meier, A. (2017). Mould machining and injection moulding of diffractive microstructures. *Journal of Manufacturing Processes, 26*, 290–294. https://doi.org/10.1016/j.jmapro.2017.02.014.

15. Aidibe, A., Nejad, M. K., Tahan, A., Jahazi, M., & Cloutier, S. G. (2016). Surface characterization of die inserts used for LED lamp plastic lenses. *The International Journal of Advanced Manufacturing Technology, 88*(9–12), 3395–3403. https://doi.org/10.1007/s00170-016-9038-x.

16. Tsai, K. M., Hsieh, C. Y., & Lo, W. C. (2009). A study of the effects of process parameters for injection moulding on surface quality of optical lenses. *Journal of Materials Processing Technology, 209*(7), 3469–3477. https://doi.org/10.1016/j.jmatprotec.2008.08.006.

17. Tsai, K. M., & Luo, H. J. (2014). An inverse model for injection moulding of optical lens using artificial neural network coupled with genetic algorithm. *Journal of Intelligent Manufacturing, 28*(2), 473–487. https://doi.org/10.1007/s10845-014-0999-z.

18. Tsai, K. M., & Lan, J. K. (2015). Correlation between runner pressure and cavity pressure within injection mould. *The International Journal of Advanced Manufacturing Technology, 79*(1–4), 273–284. https://doi.org/10.1007/s00170-014-6776-5.

19. Tsai, K. M. (2010). Effect of injection moulding process parameters on optical properties of lenses. *Applied Optics, 49*(31), 6149–6159. https://doi.org/10.1364/ao.49.006149.

20. Tsai, K. M., & Tang, B. H. (2014). Determination of injection moulding process window based on form accuracy of lens using response surface methodology. *The International Journal of Advanced Manufacturing Technology, 75*(5–8), 947–958. https://doi.org/10.1007/s00170-014-6185-9.

21. Sabiston, T., Inal, K., & Lee-Sullivan, P. (2020). Application of Artificial Neural Networks to predict fibre orientation in long fibre compression moulded composite materials. *Composites Science and Technology, 190*, 108034. https://doi.org/10.1016/j.compscitech.2020.108034.

22. Barnes, H. A. (2000). Advances in the flow and rheology of non-newtonian fluids parts A & B by D.A. Siginer, D. De Kee and R.P. Chhabra (Eds.), Elsevier, Amsterdam, 1999, 1515 pp., US$ 380, Dfl 750. *Journal of Non-Newtonian Fluid Mechanics, 91*(2–3), 298–299. https://doi.org/10.1016/S0377-0257(99)00101-9.

23. Kim, N. H., & Isayev, A. I. (2013). Birefringence in injection-compression moulding of amorphous polymers: Simulation and experiment. *Polymer Engineering & Science, 53*(8), 1786–1808. https://doi.org/10.1002/pen.23429.

24. Kwak, T. S., Suzuki, T., Baeetal, W. B., Uehara, Y., & Ohmori, H. (2005). Application of neural network and computer simulation to improve surface profile of injection moulding optic lens. *Journal of Materials Processing Technology, 170*(1–2), 24–31. https://doi.org/10.1016/j.jmatprotec.2005.04.099.

25. Loaldi, D., Calaon, M., Quagliotti, D., Parenti, P., Annoni, M., & Tosello, G. (2018). Tolerance verification of precision injection moulded Fresnel lenses. *Procedia CIRP*, *75*, 137–142. https://doi.org/10.1016/j.procir.2018.05.004.

26. Bickerton, S., & Abdullah, M. Z. (2003). Modelling and evaluation of the filling stage of injection/compression moulding. *Composites Science and Technology*, *63*(10), 1359–1375. https://doi.org/10.1016/S0266-3538(03)00022-8.

27. Schmachtenberg, E., & Johannaber, F. (2007). Montagespritzgießen–Verfahrensprinzipien und Definition. In *Technical Conference Montagespritzgießen* (pp. 1–18). Erlangen: Institute of Polymer Technology.

28. Eckardt, H. (1987). Co-injection charting new territory and opening new markets. *Journal of Cellular Plastics*, *23*(6), 555–592. https://doi.org/10.1177/0021955x8702300604.

29. Selden, R. (2000). Co-injection moulding: Effect of processing on material distribution and mechanical properties of a sandwich moulded plate. *Polymer Engineering & Science*, *40*(5), 1165–1176. https://doi.org/10.1002/pen.11244.

30. Yang, M., Wang, K., Ye, L., Mai, Y. W., & Wu, J. (2003). Low density polyethylene-polypropylene blends: Part 1 – ductility and tensile properties. *Plastics, Rubber and Composites*, *32*(1), 21–26. https://doi.org/10.1179/146580103225009112.

31. Messaoud, D., Sanchagrin, B., & Derdouri, A. (2005). Study on mechanical properties and material distribution of sandwich plaques moulded by co-injection. *Polymer Composites*, *26*(3), 265–275. https://doi.org/10.1002/pc.20084.

32. Vervoort, P., & Martens, M. (2019). In Woodhead Publishing series in metals and surface engineering. In *Handbook of Metal Injection Moulding* (2nd ed., pp. 173–194). Woodhead Publishing. https://doi.org/10.1016/B978-0-08-102152-1.00010-6.

33. He, H., Li, Y. M., & Zhang, J. G. (2010). An experimental study of metal co-injection moulding with sequential injection. *Advanced Materials Research*, *97–101*, 1116–1119. https://doi.org/10.4028/www.scientific.net/amr.97-101.1116.

34. Davie, S., Diegel, S., & Boundy, R. (2013). Transportation energy data book: Edition 32. *Energy Conservation Consumption and Utilization*, *176*(3), 319–338. https://doi.org/10.2172/1110942.

35. Yang, W., & Yokoi, H. (2003). Visual analysis of the flow behavior of core material in a fork portion of plastic sandwich injection moulding. *Polymer Testing*, *22*(1), 37–43. https://doi.org/10.1016/S0142-9418(02)00046-6.

36. Su, Q., Zhang, N., & Gilchrist, M. D. (2016). The use of variotherm systems for micro-injection moulding. *Journal of Applied Polymer Science*, *133*(9), 59–65. https://doi.org/10.1002/app.42962.

37. Giusti, R., & Lucchetta, G. (2014). Modelling the adhesion bonding mechanism in overmoulding hybrid structural parts for lightweight applications. *Key Engineering Materials*, *611*, 915–921. https://doi.org/10.4028/www.scientific.net/KEM.611-612.915.

38. Cerda, A., Oms, M., Forteza, R., & Cerdà, V. (1998). Sequential injection sandwich technique for the simultaneous determination of nitrate and nitrite. *Analytica Chimica Acta*, *371*(1), 63–71. https://doi.org/10.1016/S0003-2670(98)00278-5.

39. Baird, D., & Wilkes, G. (1983). Sandwich injection moulding of thermotropic copolyesters and filled polyester. *Polymer Engineering & Science*, *23*(11), 632–636. https://doi.org/10.1002/pen.760231108.

40. Schlatter, G., Agassant, J., Davidoff, A., & Vincent, M. (1999). An unsteady multifluid flow model: Application to sandwich injection moulding process. *Polymer Engineering & Science*, *39*(1), 78–88. https://doi.org/10.1002/pen.11398.

41. Lauritzen, J. I., & Hoffman, J. D. (1959). Formation of polymer crystals with folded chains from dilute solution. *Journal of Chemical Physics*, *31*(6), 1680–1681. https://doi.org/10.1063/1.1730678.

42. Portale, G., Cavallo, D., Alfonso, G. C., Hermida-Merino, D., Van Drongelen, M., Balzano, L., Peters, G. W. M., Goossens, J. G. P., & Bras, W. (2013). Polymer crystallization studies under processing-relevant conditions at the SAXS/WAXS DUBBLE beamline at the ESRF. *Journal of Applied Crystallography, 46*(6), 1681–1689. https://doi.org/10.1107/S0021889813027076.

43. Wang, T. H., & Young, W. B. (2005). Study on residual stresses of thin-walled injection moulding. *European Polymer Journal, 41*(10), 2511–2517. https://doi.org/10.1016/j.eurpolymj.2005.04.019.

44. Wimberger-Friedl, R., & De Bruin, J. G. (1993). Birefringence in polycarbonate: Molecular orientation induced by cooling stresses. I. Free quenching. *Journal of Polymer Science: Polymer Physics, 31*(8), 1041–1049. https://doi.org/10.1002/polb.1993.090310814.

45. Wimberger-Friedl, R. (1995). The assessment of orientation, stress and density distributions in injection-moulded amorphous polymers by optical techniques. *Progress in Polymer Science, 20*(3), 369–401. https://doi.org/10.1016/0079-6700(94)00036-2.

46. Yao, D., & Kim, B. (2002). Increasing flow length in thin wall injection moulding using a rapidly heated mould. *Polymer-Plastics Technology and Engineering, 41*(5), 819–832. https://doi.org/10.1081/PPT-120014390.

47. Yang, C., Su, L., Huang, C., Huang, H. X., Castro, J. M., & Yi, A. Y. (2011). *Adv. Polym. Tech., 30*(1), 51–61. https://doi.org/10.1002/adv.20211.

48. Yu, L., Koh, C. G., James Lee, L., Koelling, K. W., & Madou, M. J. (2002). *Polymer Engineering & Science, 42*(5), 871–888. https://doi.org/10.1002/pen.10998.

49. Chivatanasoontorn, V., Yamada, K., & Kotaki, M. (2015). Highly oriented microstructures and surface mechanical properties of polypropylene (PP) moulded by ultra-high shear rate. *Polymer, 72*, 104–112. https://doi.org/10.1016/j.polymer.2015.07.013.

50. Min, I., & Yoon, K. (2011). An experimental study on the effects of injection-moulding types for the birefringence distribution in polycarbonate discs. *Korea-Australia Rheology Journal, 23*(3), 155–162. https://doi.org/10.1007/s13367-011-0019-1.

51. Weng, C., Lee, W. B., To, S., & Jiang, B. Y. (2009). Numerical simulation of residual stress and birefringence in the precision injection moulding of plastic microlens arrays. *International Communications in Heat and Mass Transfer, 36*(3), 213–219. https://doi.org/10.1016/j.icheatmasstransfer.2008.11.002.

52. Weng, C., Lee, W. B., & To, S. (2009). Birefringence techniques for the characterization of residual stresses in injection-moulded micro-lens arrays. *Polymer Testing, 28*(7), 709–714. https://doi.org/10.1016/j.polymertesting.2009.06.007.

53. Lee, Y. B., Kwon, T. H., & Yoon, K. (2002). Numerical prediction of residual stresses and birefringence in injection/compression moulded center-gated disk. Part I: Basic modelling and results for injection moulding. *Polymer Engineering & Science, 42*(11), 2246–2272. https://doi.org/10.1002/pen.11114.

54. Isayev, A. I., Shyu, G. D., & Li, C. T. (2006). Residual stresses and birefringence in injection moulding of amorphous polymers: Simulation and comparison with experiment. *Journal of Polymer Science: Polymer Physics, 44*(3), 622–639. https://doi.org/10.1002/polb.20724.

55. Pantani, R., Coccorullo, I., Speranza, V., & Titomanlio, G. (2007). Morphology evolution during injection moulding: Effect of packing pressure. *Polymer, 48*(9), 2778–2790. https://doi.org/10.1016/j.polymer.2007.03.007.

56. Meekers, I., Refalo, P., & Rochman, A. (2018). Analysis of process parameters affecting energy consumption in plastic injection moulding. *Procedia CIRP, 69*, 342–347. https://doi.org/10.1016/j.procir.2017.11.042.

57. Lucchetta, G., Masato, D., & Sorgato, M. (2018). Optimization of mould thermal control for minimum energy consumption in injection moulding of polypropylene parts. *Journal of Cleaner Production, 182*, 217–226. https://doi.org/10.1016/j.jclepro.2018.01.258.

58. Yang, S. Y., & Nien, L. (2006). *Advances in Polymer Technology, 15*(3), 205–213. https://doi.org/10.1002/adv.1996.060150302.

59. Chang, C. Y. (2006). Simulation of mould filling in simultaneous resin injection/compression moulding. *Journal of Reinforced Plastics and Composites*, *25*(12), 1255–1268. https://doi.org/10.1177/0731684406060253.

60. Lee, B. H., & Kim, B. H. (1996). Automated design for the runner system of injection moulds based on packing simulation. *Polymer-Plastics Technology and Engineering*, *35*(1), 147–168. https://doi.org/10.1080/03602559608000086.

61. Christiansen, A. B., Clausen, J. S., Mortensen, N. A., & Kristensen, A. (2014). Injection moulding antireflective nanostructures. *Microelectronic Engineering*, *121*(6), 47–50. https://doi.org/10.1016/j.mee.2014.03.027.

62. Lucchetta, G., Sorgato, M., Zanchetta, E., Brusatin, G., Guidi, E., Di Liddo, R., & Conconi, M. T. (2015). Effect of injection molded micro-structured polystyrene surfaces on proliferation of MC3T3-E1 cells. *Express Polymer Letters*, *9*(4), 354–361. https://doi.org/10.3144/expresspolymlett.2015.33

63. Sorgato, M., Masato, D., & Lucchetta, G. (2017). Effect of vacuum venting and mould wettability on the replication of micro-structured surfaces. *Microsystem Technologies*, *23*, 2543–2552. https://doi.org/10.1007/s00542-016-3038-5.

64. Guo, F., Zhou, X., Liu, J., Zhang, Y., Li, D., & Zhou, H. (2019). A reinforcement learning decision model for online process parameters optimization from offline data in injection moulding. *Applied Soft Computing Journal*, *85*, 105828. https://doi.org/10.1016/j.asoc.2019.105828.

65. Zhao, J., Wang, G., Zhang, L., Li, B., Wang, C., Zhao, G., & Park, C. B. (2019). Lightweight and strong fibrillary PTFE reinforced polypropylene composite foams fabricated by foam injection moulding. *European Polymer Journal*, *119*, 22–31. https://doi.org/10.1016/j.eurpolymj.2019.07.016.

66. Einarsdottir, E. R., Geminiani, A., Chochlidakis, K., Feng, C., Tsigarida, A., & Ercoli, C. (2019). Dimensional stability of double-processed complete denture bases fabricated with compression moulding, injection moulding, and CAD-CAM subtraction milling. *Journal of Prosthetic Dentistry*, *124*, 116–121. https://doi.org/10.1016/j.prosdent.2019.09.011.

67. Guan, B., Cherrill, M., Pai, J. H., & Priest, C. (2019). Effect of mould roughness on injection moulded poly (methyl methacrylate) surfaces: Roughness and wettability. *Journal of Manufacturing Processes*, *48*, 313–319. https://doi.org/10.1016/j.jmapro.2019.10.024.

68. Yan, X., Yang, Y., & Hamada, H. (2018). Tensile properties of glass fibre reinforced polypropylene composite and its carbon fibre hybrid composite fabricated by direct fibre feeding injection moulding process. *Polymer Composites*, *39*(10), 3564–3574. https://doi.org/10.1002/pc.24378.

69. Lu, Y., Jiang, K., Y. Liu, Zhang, Y., & Wang, M. (2020). Study on mechanical properties of co-injection self-reinforced single polymer composites based on micro-morphology under different moulding parameters. *Polymer Testing*, *83*, 106306. https://doi.org/10.1016/j.polymertesting.2019.106306.

70. Manikandan, R., Arjunan, T. V., & Nath, O. P. A. R. (2019). Studies on micro structural characteristics, mechanical and tribological behaviours of boron carbide and cow dung ash reinforced aluminium (Al 7075) hybrid metal matrix composite. *Composites Part B: Engineering*, *183*, 107668. https://doi.org/10.1016/j.compositesb.2019.107668.

71. Tosello, G., & Costa, F. S. (2019). High precision validation of micro injection moulding process simulations. *Journal of Manufacturing Processes*, *48*, 236–248. https://doi.org/10.1016/j.jmapro.2019.10.014.

72. Wang, Q., Li, X., Chang, T., Hu, Q., & Yang, X. (2018). Terahertz spectroscopic study of aeronautical composite matrix resins with different dielectric properties. *Optik*, *168*, 101–111. https://doi.org/10.1016/j.ijleo.2018.04.019.

73. Kumar Yadav, R., Hasan, Z., & Husain Ansari, A. (2019). Investigation of mechanical and wear behavior of Al based SiC reinforce metal matrix composite. *Materials Today: Proceedings*, *21*, 1537–1543. https://doi.org/10.1016/j.matpr.2019.11.083.

74. Natrayan, L., & Senthil Kumar, M. (2019). Optimization of wear behaviour on AA6061/ Al_2O_3/SiC metal matrix composite using squeeze casting technique – statistical analysis. *Materials Today: Proceedings*, *27*, 306–310. https://doi.org/10.1016/j.matpr.2019.11.038.

75. Guo, Q., Han, Y., & Zhang, D. (2020). Interface-dominated mechanical behavior in advanced metal matrix composites. *Nano Materials Science*, *2*, 66–71. https://doi.org/10.1016/j.nanoms.2020.03.007.

76. Joo, Y. A., Kim, Y. K., Yoon, T. S., & Lee, K. A. (2018). Microstructure and high temperature oxidation property of Fe-Cr-B based metal/ceramic composite manufactured by powder injection moulding process. *Metals and Materials International*, *24*(2), 371–379. https://doi.org/10.1007/s12540-018-0053-3.

77. Zhang, Z., Chen, X., Zhang, G., & Feng, C. (2020). Synthesis of MoO_3/V_2O_5/C composite as novel anode for Li-Ion battery application. *Journal of Nanoscience and Nanotechnology*, *20*(5), 2911–2916. https://doi.org/10.1166/jnn.2020.17441.

78. Lu, G., Liu, X., Zhang, P., Bao, L., & Zhao, B. (2020). Nanoarchitectonic composites of mixed and covalently linked multiwalled carbon nanotubes and tetra-[α-(p-amino)benzyloxyl] phthalocyanine zinc(II). *Journal of Nanoscience and Nanotechnology*, *20*(5), 2713–2721. https://doi.org/10.1166/jnn.2020.17472.

79. Ryutaro, W., & Tatsuo, K. (2020). Surfactant-assisted mesostructural variation by the molecular structure of frameworks. *Journal of Nanoscience and Nanotechnology*, *20*(5), 3078–3083. https://doi.org/10.1166/jnn.2020.17478.

80. Ling, Q., Wei, J., Chen, L., Zhao, H., Lei, Z., Zhao, Z., Xie, R., Ke, Q., & Cui, P. (2020). Solvothermal synthesis of humic acid-supported CeO_2 nanosheets composite as high performance adsorbent for congo red removal. *Journal of Nanoscience and Nanotechnology*, *20*(5), 3225–3230. https://doi.org/10.1166/jnn.2020.17384.

81. Robertsam, A., & Victor Jaya, N. (2020). Fabrication of a low-coercivity, large-magnetoresistance PVA/Fe/Co/Ni nanofibre composite using an electrospinning technique and its characterization. *Journal of Nanoscience and Nanotechnology*, *20*(6), 3504–3511. https://doi.org/10.1166/jnn.2020.17404.

82. Zhang, J., Wu, X., Tong, X., Zhang, C., Wang, H., Su, J., Jia, Y., Zhang, M., Chang, T., & Fu, Y. (2020). $AgBrO_3$/Few-Layer g-C_3N_4 composites: A visible-light-driven photocatalyst for tetracycline degradation. *Journal of Nanoscience and Nanotechnology*, *20*(6), 3424–3431. https://doi.org/10.1166/jnn.2020.17488.

83. Shen, S., Kanbur, B. B., Zhou, Y., & Duan, F. (2019). Thermal and mechanical analysis for conformal cooling channel in plastic injection moulding. *Materials Today: Proceedings*, *28*, 396–401. https://doi.org/10.1016/j.matpr.2019.10.020.

84. Zink, B., Kovács, N. K., & Kovács, J. G. (2019). Thermal analysis based method development for novel rapid tooling applications. *International Communications in Heat and Mass Transfer*, *108*, 104297. https://doi.org/10.1016/j.icheatmasstransfer.2019.104297.

85. Shamsuzzaman, M., Haridy, S., Maged, A., & Alsyouf, I. (2019). Design and application of dual-EWMA scheme for anomaly detection in injection moulding process. *Computers & Industrial Engineering*, *138*, 106132. https://doi.org/10.1016/j.cie.2019.106132.

86. Liu, H., Zhang, X., Quan, L., & Zhang, H. (2020). Research on energy consumption of injection moulding machine driven by five different types of electro-hydraulic power units. *Journal of Cleaner Production*, *242*, 118355. https://doi.org/10.1016/j.jclepro.2019.118355.

87. Cao, W., Shen, Y., Wang, P., Yang, H., Zhao, S., & Shen, C. (2019). Viscoelastic modelling and simulation for polymer melt flow in injection/compression moulding. *Journal of Non-Newtonian Fluid Mechanics*, *274*, 104186. https://doi.org/10.1016/j.jnnfm.2019.104186.

88. Liu, H., Pei, C., Yang, J., & Yang, Z. (2020). Influence of long-term thermal aging on the microstructural and tensile properties of all-oxide ceramic matrix composites. *Ceramics International*, *46*(9), 13989–13996. https://doi.org/10.1016/j.ceramint.2020.02.198.

89. Gong, J., Miao, H., & Zhao, Z. (2001). The influence of TiC-particle-size on the fracture toughness of Al_2O_3–30 wt.%TiC composites. *Journal of the European Ceramic Society*, *21*(13), 2377–2381. https://doi.org/10.1016/s0955-2219(01)00206-0.

90. Peillon, F. C., & Thevenot, F. (2002). Microstructural designing of silicon nitride related to toughness. *Journal of the European Ceramic Society, 22*(3), 271–278. https://doi.org/10.1016/s0955-2219(01)00290-4.

91. Deng, L., Qiao, L., Zheng, J., Ying, Y., Yu, J., Li, W., Che, S., & Cai, W. (2019). Injection moulding, debinding and sintering of ZrO_2 ceramic modified by silane couping agent. *Journal of the European Ceramic Society, 40*(4), 1566–1573. https://doi.org/10.1016/j.jeurceramsoc.2019.11.069.

92. Wang, G., Zhao, J., Wang, G., Zhao, H., Lin, J., Zhao, G., & Park, C. (2020). Strong and super thermally insulating in-situ nanofibrillar PLA/PET composite foam fabricated by high-pressure microcellular injection moulding. *Chemical Engineering Journal, 390*, 124520. https://doi.org/10.1016/j.cej.2020.124520.

93. Lee, R. E., Azdast, T., Wang, G., Wang, X., Lee, P. C., & Park, C. B. (2020). Highly expanded fine-cell foam of polylactide/polyhydroxyalkanoate/nano-fibrillated polytetrafluoroethylene composites blown with mould-opening injection moulding. *International Journal of Biological Macromolecules, 155*, 286–292. https://doi.org/10.1016/j.ijbiomac.2020.03.212.

94. Wang, M. W., Arifin, F., & Huynh, T. T. N. (2019). Optimization of moulding parameters for a micro gear with Taguchi method. *Journal of Physics: Conference Series, 1167*, 012001. https://doi.org/10.1088/1742-6596/1167/1/012001.

95. Zhai, W., Sun, R., Sun, H., Ren, M., Dai, K., Zheng, G., Liu, C., & Shen, C. (2018). Segregated conductive CNTs/HDPE/UHMWPE composites fabricated by plunger type injection moulding. *Materials Letters, 229*, 13–16. https://doi.org/10.1016/j.matlet.2018.06.114.

96. Binet, C., Heaney, D. F., Spina, R., & Tricarico, L. (2005). Experimental and numerical analysis of metal injection moulded products. *Journal of Materials Processing Technology, 164–165*, 1160–1166. https://doi.org/10.1016/j.jmatprotec.2005.02.128.

97. Zhang, H. L., Ong, N. S., & Lam, Y. C. (2008). Mould surface roughness effects on cavity filling of polymer melt in micro injection moulding. *The International Journal of Advanced Manufacturing Technology, 37*(11–12), 1105–1112. https://doi.org/10.1007/s00170-007-1060-6.

98. Wang, G., Zhao, G., Li, H., & Guan, Y. (2009). Research on a new variotherm injection moulding technology and its application on the moulding of a large LCD panel. *Polymer-Plastics Technology and Engineering, 48*(7), 671–681. https://doi.org/10.1080/03602550902824549.

99. Ahmad, F. (2005). Orientation of short fibres in powder injection moulded aluminum matrix composites. *Journal of Materials Processing Technology, 169*(2), 263–269. https://doi.org/10.1016/j.jmatprotec.2005.03.036.

100. Arao, Y., Fujiura, T., Itani, S., & Tanaka, T. (2015). Strength improvement in injection-moulded jute-fibre-reinforced polylactide green-composites. *Composites Part B: Engineering, 68*, 200–206. https://doi.org/10.1016/j.compositesb.2014.08.032.

101. Sorgato, M., Babenko, M., Lucchetta, G., & Whiteside, B. (2017). Investigation of the influence of vacuum venting on mould surface temperature in micro injection moulding. *The International Journal of Advanced Manufacturing Technology, 88*, 547–555. https://doi.org/10.1007/s00170-016-8789-8.

102. Fu, H., Xu, H., Liu, Y., Yang, Z., Kormakov, S., Wu, D., & Sun, J. (2020). Overview of injection moulding technology for processing polymers and their composites. *ES Materials' Manufacturing.* https://doi.org/10.30919/esmm5f713

103. Sztorch, B., Brząkalski, D., Jałbrzykowski, M., & Przekop, R. E. (2021). Processing technologies for crisis response on the example of COVID-19 pandemic—injection moulding and FFF case study. *Processes, 9*, 791. https://doi.org/10.3390/pr9050791

8 Futuristic 3D Printing Applications

8.1. 3D PRINTING IN CLASSROOM

In both educational and professional settings, assistive technology (AT) can increase the independence of young adults with ID [1]. It can help students while they work on critical job-related competencies like budgeting and independent travel [2]. Supporting activities of daily living (ADLs) and task completion are frequently examined in this area of research as a skill for employment. Bauchet et al. developed "Archipel," a system that tracks task completion for ADLs, and tested it on people with ID [3]. Their system might take advantage of input from sensors or direct input, including interacting with prompts on a touch screen. In two instances of research, Chang et al. investigated prompting systems for people with cognitive impairment. Inside one study, they used engagement prompts to recognise task completion in a professional setting using sensor nodes, Bluetooth connectivity, and a PDA [4]. Later, they developed Kinempt, a similar device based on the Microsoft Kinect [5]. There are a few examples of studies that examine how technology is used and/or learned in relation to ID. Two investigations with an emphasis on technology use by people with ID were carried out by Feng and colleagues. In their 2010 publication, they describe the results of questionnaires given to families of children with ID, and they list the main difficulties these kids have when using computers from the perspective of their parents.

In a related study, they made ethnographic observations of working people with ID [6]. They describe how professionals use office software, communication tools, and security solutions like CAPCHA for this based on empirical data.

Dunn et al. identified six special education hurdles that prevent people with ID from having the opportunity to learn about technology and related areas [7]: Low participation rates in both structured and unstructured STEM-related activities are caused by a lack of STEM role models, parent and teacher misconceptions, a lack of appropriate information and counselling, educators' lack of knowledge and expertise in including ID students, technical obstacles, and other factors [8]. These impediments on student opportunities to explore STEM disciplines are due to inadequate expectations of students with disabilities' skills. Our study expands on the exploratory work that looked at how people with intellectual disabilities used technology, and we offer new insight into how to teach an attempt to cut technology to young adults with ID.

Education and Intellectual Disability It is a widely held perception in society that postsecondary education expands more opportunities for employment and the potential for greater income. In a study of students with ID, 58% of all students who pursued higher education got jobs after completing their studies. In contrast, 32% of individuals with IDs did not enroll in postsecondary education. Poor career-related decisions could be the consequence of a majority of these individuals' lack of

DOI: 10.1201/9781003349341-8

self-determination and comprehension of the decision-making approach [9]. Students with ID are afforded greater practice and patience while making career-related decisions in postsecondary education, which directly improves their confidence and reduces the negative emotions that also can obstruct outcome. Our framework of this study on ID and special education is supported by this contribution. We emphasise the educational process in relation to 3D printing, a rapidly evolving technology that is typically regarded as being technical and engineering-heavy and perhaps "beyond" the level of our students.

Special Populations and 3D Printing The physical aspects of 3D printed objects have increased the popularity of this technology.

Applications of 3D printed visuals for students who are visually impaired have been investigated in recent research. In the context of math teaching, Brown and Hurst [10] introduced VizTouch, an algorithm efficient for generating 3D printable line graphs. More recently, Braier, et al. [11] evaluated the prospect of using 3D printed artwork to enable inclusive education and demonstrated their own software implementation for optimizing the fabrication of 3D printable haptic aid. The use of 3D printed documentation as just a resource of computer science education and STEM stimulation for young people with visual impairments was also reviewed by Kane and Bigham. These ventures generally focus on generating 3D prints for people with disability.

Another emerging field of research is the direct use of 3D printers by individuals who are disabled. The advantages and disadvantages of 3D printers in special education classrooms were discussed by Buehler et al. [12]. Hook et al. outlined obstacles for adolescent individuals who had disabilities in generating 3D printed assistive technology (AT) devices in a more comprehensive exploration of fabrication methods for DIY assistive technology. Buehler et al. noticed that a minority of designers with disabilities shared their DIY-AT on the website Thingiverse1 when investigating 3D printable AT. We are able to learn more about the ways persons with disabilities may use 3D printing technology in their education and daily life by working closely with students with ID. We discovered a variety of instructional techniques used by instructors to introduce students to 3D printing, such as simple beginning projects for a pleasant exposure to design and precise guidelines for more ambitious projects. The instructors also detected fluctuations in teenage folks' innovation activities based on whether they were obliged to accomplish specific architectural objectives or encouraged to construct improvisation objects. According to the instructors, starting students off on their path to achievement from the beginning is the most essential approach to introducing 3D printing to adolescents. Youth must be safeguarded against failure, especially in the initial stages of learning, according to respondents. They noted that complete guidance can cause annoyance and obstacles. Instructors who provided their students more creative projects also made more allusions to the alleged excitement of their students. Although open-ended lessons might be more interesting, we were cautioned that they can be difficult to coordinate with more students and demands. We constructed a variety of wide and organized lessons based on these experiences. While we frequently provided our students the accountability for the creative design elements, we also afforded them structure in the way of strategic planning and parameters for each project [2].

The connection between technology and methodological and material knowledge is one of the most fiercely debated topics in teacher education for employing digital technologies. The ICT courses of teacher preparation programmes frequently feature stand-alone technology courses (Bakir, 2015; Park, Jun, & Kim, 2011; Tondeur et al., 2012), but research indicates that integrating technology into course content and pedagogy is a more viable strategy (Admiraal et al., 2017; Angeli & Valanides, 2009; Bakir, 2015; So & Kim, 2009; Voogt, Fisser, Pareja Roblin, Tondeur, & van Braak, 2013). Few studies have distinguished emerging technologies from the technologies that have become ubiquitous when defining technological knowledge, despite the extensive research on how to prepare, which was incorporate ICT technologies into K–12 instruction (Chai, Koh, & Tsai, 2013; Kaufman, 2014; Kay, 2006). (Cox & Graham, 2009). Studies on using new technologies, like VR, frequently involve teachers who are already adept in the necessary skills and expertise to start technology-infused teaching methods. However, the introduction of new technology to pre-service teacher preparation and ongoing professional development for teachers has received less attention. TPACK, a conceptual framework for technology education, based on Shulman's description of "pedagogical content knowledge," was introduced by Mishra and Koehler in 2006 (Figure 8.1) (Shulman, 1986). The use of the Chosen methodology in methods of teaching and learning has rapidly spread to include professional growth and classroom technology incorporation. It has also been utilized to assess and enhance, which was before basic educational programmes across a variety of subjects and levels (Admiraal et al., 2017; Chai, Koh, & Tsai, 2010; Voogt et al., 2013; Voogt & McKenney, 2017) (Figure 8.1. TPACK framework. Source: Reproduced by permission of the publisher, © tpack.org).

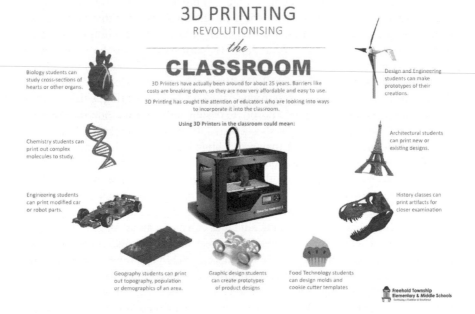

FIGURE 8.1   3D Printing in the Classroom.

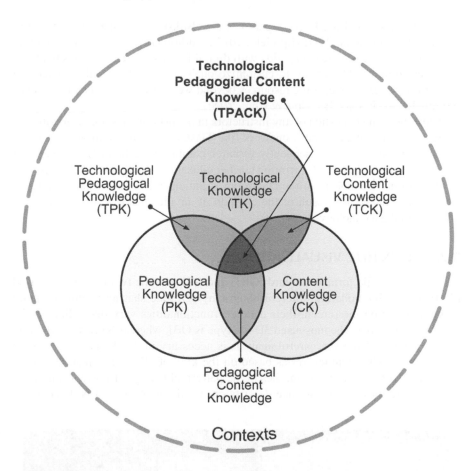

FIGURE 8.2   Technological Pedagogical Content Knowledge Framework [9].

Research shows that recent studies investigating the ICT integration in classrooms from the framework of TPACK involved: (1) subject-general technologies including mainly multiple ICT tools, web-based technology, and movie-making software; (2) subject-specific technologies in science, engineering, and mathematics subjects including specialized technologies such as graphing calculators and simulation, digital microscopes, etc [2]. The educational potential of 3D printing, in contrast to prior innovations, is based on the concept that it integrates the virtual and physical worlds, permitting the technology to be used in both specific topic and particular matter context. A 3D model can be generated on the computer screen for aesthetic purposes utilizing 3D printing technology. By combining it with various methods such as the Internet of Everything or more traditional techniques like painting, it can also be utilised to create anything other than a 3D-printed plastic object. In this way, flipped classrooms, hands-on learning, and cross-disciplinary courses can all benefit from digital fabrication technologies. Media outlets have reported on a large number of successful cases, and academic papers to date (Blikstein, 2013; Bull et al., 2010; Halverson & Sheridan, 2014; Highfield, 2015). Additionally, studies addressed that applications in creating teaching aids with digital fabrication technologies hold further opportunities. The ability to customize one's own teaching aids provides an opportunity to expand teaching resources beyond web-based and screen-based ones (Highfield, *2015*; Song, Ha, Goo, & Cho, *2018*).

The continuing education of any instructors in 3D printing has not been critically evaluated using a conceptual framework like TPACK, despite the growing body of research on the subject. Consequently, research comprising a thorough evaluation of the current curriculum for 3D printing teacher education is required. The integration of 3D printing technology in the classroom is anticipated to undertake prolonged, widespread change as a result of the creation of an adequate teacher training programme in the technology [1].

8.2. SCIENTIFIC VISUALISATION

The Wavefront file format standard (OBJ) and STL are the two most widely used file types for 3D modelling (StereoLithography). Despite sharing a common point of view in 3D on the screen, their detailed functionalities vary depending on the development's goal. The most used 3D file type is OBJ, whereas STL is mostly used for 3D printing. However, careful analysis is necessary in the domain of scientific visualisation. The accuracy varies based on the kind of 3D data format. Given the fact that OBJ and STL create meshes in the pattern of triangular polygons, they are not adequate for excellent surface description. Furthermore, it will be finer if you

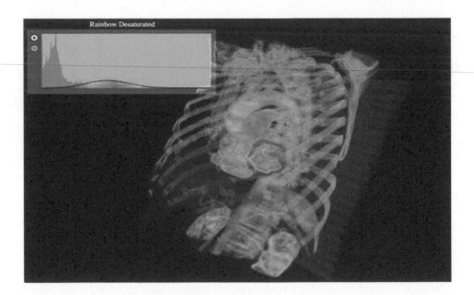

FIGURE 8.3   Scientific Visualization Using VTK Files [3].

```
# vtk DataFile Version 2.0              ](1)
Really cool data                      ](2)
ASCII | BINARY                  ](3)
DATASET type          ](4)
...
POINT_DATA n                        ](5)
...
CELL_DATA n
```

<VTKFile type="ImageData" version="0.1"
byte_order="LittleEndian">
...
</VTKFile>

(a)   (b)

Figure 2. Two Types of VTK files (a) simple legacy Type (b) XML Type

FIGURE 8.4 Two Types of VTK Files (a) Simple Legacy Type (b) XML Type [3].

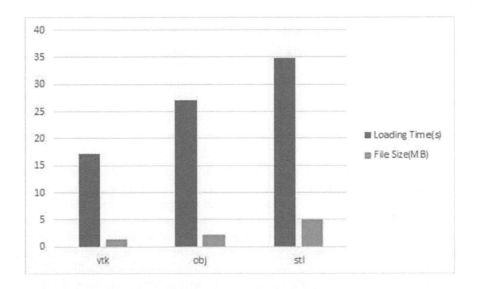

FIGURE 8.5 Performance Comparison on the Web [3].

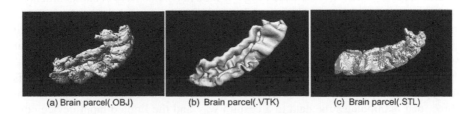

(a) Brain parcel(.OBJ)   (b) Brain parcel(.VTK)   (c) Brain parcel(.STL)

FIGURE 8.6 Brain Parcel Implementation (a) OBJ (b) VTK (c) STL [3].

TABLE 8.1

Experiment Environment.

Classification		Information	
3D File Format	VTK	OBJ	STL
Material (Color)		MeshLambertMaterial (0xffffff)	
Side		DoubleSide	
3D JavaScript Library		Three.js	
Used JavaScript Library		Vanilla.js (Pure JavaScript)	
Light (Color)		DirectionalLight (0xffffff)	

TABLE 8.2

Performance Comparison on the Web.

Classification	VTK	OBJ	STL
3D File Format		Chrome v.72.0.3626.109	
Material (Color)	17.18	17.18	17.18
Side	1.4	2.2	4.9
Network Environment		Fast 3G	

expand the number of triangle polygons, but doing so geometrically expands the file size and requires a lot of CPU power. In contrast, VTK offers a selection of polygon structures, including cube, rectangular, and triangular polygons. This makes it feasible to show fragile textures. It is feasible to render delicate surfaces, and the file size is minimal. The conception and construction of VTK are explained in this article. In the Chrome browser, they also observed the processing durations and file sizes for VTK, STL, and OBJ. Moreover, the screen where each 3D file format is exploited under identical conditions allows the user to intuitively view the differences in surface rendering capability between VTK, STL, and OBJ. This project is anticipated to assist in the creation of a 3D shape format for use in web-based scientific visualisation [3].

Virtual reality (VR) research has advanced significantly throughout the years, and several success stories from a wide range of industry fields have been documented. Without a question, the user's improved sense of presence and the convenience of involvement with the fundamental scientific evidence are indeed the greatest benefit of using VR for scientific visualisation. By merging the application with the familiar real world, augmented reality (AR) techniques further optimises these. For interacting with scientific data, two-dimensional (2D) and three-dimensional (3D) widgets are typically employed [13]. These widgets operate as an interface between a virtual object that manipulates the data in a certain way and the limited input device controlled by the users. Pressing a button on the input device initiates an action on a widget. There are large collections of 2D and 3D widgets that have been developed for an array of activities. 2D widgets, for illustration, can be used to modify simulation or visualisation parameters. You can use 3D widgets to query dataset values, define seed points for flow visualisation approaches, and other things. When it comes to input devices, the two degrees of freedom (2DOF) mouse, six degrees of freedom (6DOF) devices (such as wands, gloves, etc.), and haptic devices have historically been the most common options. The devices have buttons that can be pressed to start actions. But in every scenario, a point region in the 3D interaction area is regulated by these gadgets. We believe that providing users only pointing devices contributes to unintuitive and often unpleasant interfaces. In our opinion, the principal reason for this is that the user must count on the presented visual or acoustic evidence to assess the outcome of their activities, because the physical devices simply work as a proxy for a virtual item. The real interaction takes place outside of the user's familiar environment, in a remotely managed, immaterial virtual environment. Despite the utilization of sophisticated haptic devices, attempting to choose destinations out from thin air while still being prompted mostly by mediocre sensory systems is an unproductive method of interacting. Accurate, substantial tools are provided via realistic user interface design to be manipulated by the user, as well as synthetic reproductions of these props. The physical input props are exactly the same as their virtual equivalents, or at least have a similar shape, and can be handled in the same way. The user interacts by carrying out the exact same actions on the physical object as he would carry out on the virtual representation. For instance, a physical, tangible handle can be used to manipulate a virtual object.

Since it taps into the same characteristics that its user requires to sense and manipulate his external conditions in regular living, such direct manipulation of actual things offers a very natural and user-friendly interface [14]. In a sense, the user can actually grasp the data in his hands and engage with it directly rather than merely manipulating it indirectly on the screen if the prop is a visual representation of the information in a graphic display. The only physical constraints on tangible interfaces are those imposed by the types of manipulations that may be felt and recorded. Dials, sliders, buttons, switches, position and orientation tracking, pressure and tension sensors, or any other type of sensing and strain gauge may be included in the interface. Even proactive haptic feedback systems might be incorporated with the interface. In our methodology, the objects are handled with utilising two hands and a pen. The interface employs touch between the stylus and the prop as the trigger rather than buttons to indicate user actions. When this contact happens, this could be recognized using information from the monitoring system and exact calibration of the prop. This makes it possible for the interaction to take place in the same way as commonplace instruments are used: the stylus is simply picked up and used on the prop by contact. In this article, we investigate how a better interface for scientific visualisations might be created using actual input objects. Our method involves using fast prototyping tools to create three-dimensional physical items of any shape, which are then imported into the environment and used as tactile interfaces. The method is iterative, allowing for the usage of increasing complexity props as more knowledge of the fundamental simulation is achieved. We provide a summary of the related work in the section that follows. Finally, we describe our example application, in which we employ printed tangible props for interactive measurement of marine coral data. Following an explanation of the concept and its potential applications, we look at the technical requirements of systems using tangible props. Numerous systems for scientific visualisation have been created, and they accept input from a wide range of physical objects. These systems either use specific props that are specific to one specific application or kind of data, or very general abstract props. Physical props are often employed in these systems for their similarity to the virtual model and comfort when touched, whereas interaction is typically done with virtual devices and props. Specialized tools like the Phantom, to which props are occasionally fastened, provide haptic sensation. In such a configuration, mixed props are frequently employed, where just the portion handled by the user is a genuine prop and the rest of the item only exists in the virtual realm. In order to establish a natural interface for a neurosurgical visualisation tool, Hinckley et al. tracked passive props [15]. A stylus was used to guide a light into the patient's brain, and a rubber sphere was utilised to direct movement of the brain model. The point where the ray and the brain model met to form the trajectory to remove a portion of the brain, a cutting plane was controlled by a third prop made of a tracked piece of plexiglass. To create a more realistic prop, a doll's head was used in place of the sphere, however, this gave consumers false expectations because the doll's head prop did not precisely match the brain data. The virtual tool would not be positioned at a highly specific location on the head prop, such as the eyes, if the trajectory or cutting plane were set there. This suggests that props should either be highly abstract or a perfect match of the data. For a geoscience data analysis application, Couture et al. developed a tabletop projection system using physical objects as input devices [16]. To choose cutting planes across a 3D volumetric model created from seismic data, one or more physical props of various types were moved around a flat projection. It's possible to employ multiple props at once. To start various actions, a real box with physical buttons was employed. Each of these propellers represented the virtual handle of a cutting plane; as a result, the prop only served as a handle for the real tool. Any connection between the props or their physical shapes had no special significance. Virtual objects were controlled by physical ("graspable") items in irregularly shaped physical handles called "bricks." Ishii and Ullmer took this concept further with Tangible Bits, employing a more direct correspondence between physical and virtual objects in a context where almost every part of the environment could be used for interaction [6]. However, the actual props used in this concept serve as physical handles for virtual tools, or are physical user interface elements. The data with which the interaction takes place only exists in the digital domain, and has no matching physical prop (236 Virtual Reality (2009) 13:235–244 123) Gillet et al. have used three-dimensional printed props of molecules in an AR application [17]. The printed props included attachment surfaces for ARToolKit markers, which provided an implicit calibration of the props. Multiple ARToolKit markers were used for each prop, also to improve accuracy but mainly to prevent loss of tracking due to marker occlusion. The props were used to position and to orient the virtual models and were augmented with various kinds of information and visualizations, but no other input devices were used to interact with these props. As such, the props served only as physical handles for the virtual molecular models. Ortega and Coquillart created an immersive industrial automotive application with a mixed tangible prop, attached to a string-based haptic system [18]. The handle is a physical tangible prop, while the rest of the mixed prop existed only in the virtual world. The mixed prop is explicitly used to avoid calibration errors from using a full real prop. The focus is on the virtual part of the prop, so any mismatch between the real prop and the virtual model is of little consequence. The prop serves only as a physical handle for a virtual tool, and no physical form is used for the data, other than the haptic feedback. Kok and van Liere (2004) controlled 2D widgets with 3D tangible props [19]. They explored the use of passive haptic feedback as a means to detect that an action can be performed on a widget. The widgets were placed on a virtual cube, which was co-located with a tracked cube prop, and a tracked stylus was used to select the widget, but to actually start the interaction, a button had to be pressed, thus the tactile feedback from the surface of the cube only served to indicate that the stylus could interact with one of the widgets on the cube. The props only served as physical and mixed-mode user interface elements. In our previous research (Kruszynski and van Liere 2008), we had to use quantifiable interaction with such a monitored touchscreen on a tracked cube prop to control 3D widgets; however, the interaction was not modeless but rather elicited by pressing a foot pedal, it was carried out on only one surface of the cube, and the cube prop was not an exact replica of the virtual counterpart [20]. Additionally, the tangible cube prop was only a physical component of the user interface, and had nothing to do with the dataset that was the subject of the interaction. These systems all make use of physical objects as input methods. The props do not often symbolize the data; generally, they behave as abstract physical handles or tools. The option of interacting through touch is expanded by the introduction of tangible props. Since individuals are used to handling actual objects in their daily lives, holding a tangible representation of a data set in your hand provides a very natural manner of interacting. The ability to produce any physical data representation is made possible by rapid prototyping technologies. When combined with a tracking system, this printed object can then be utilised as a real-world input prop. The usage of printed tangible props is better suited for some applications than for others. This method is appropriate for applications that entail the creation or assessment of mathematical or geometrical structures or models.

Model construction iteratively,

In an iterative process, models are created: A preliminary, basic model in its first iteration. The model can then be used to simulate scenarios or do other calculations, or it can be examined. The next iteration of the model is then created once the results have been analysed and the model has been modified where necessary. The method is carried out until the finished model is thought to be satisfactory. Figure 8.1 shows this procedure. Making a real-world replica of the model for each iteration is a logical extra step in the process. A model's physical representation can provide more clarity on its characteristics. The representation can be produced using a variety of 3D printing techniques, which can automatically turn the model into a solid physical thing. It is very conceivable that 3D printers will be as common as ordinary printers in a few years and produced 3D objects will become disposable single use items as this technology becomes more and more accessible. The print can be utilised as a tangible input prop that precisely matches the virtual model if the position and orientation of the printed model are tracked and matched to the original virtual model. The virtual model that uses the touch sensation can then be inspected at that point (Figure 8.3).

The fundamental steps of model refinement are following. A basic model is made, then iteratively improved. This is seen in Figure 8.4. When examining small features, the visualisation system can display an enlarged version of the model or offer extra visualisations and information that are not apparent on the actual prop. Thus, the prop functions as both an input and an output. This view of the model's appearance is the outcome of the most recent modelling iteration, and it can be augmented in an augmented reality tool. Additionally, it serves as a tool for interacting with the model. Most of the time, virtual objects are used for interaction, and a prop can only operate as a proxy for a virtual input object. However, by including physical input devices, the interaction actually occurs in the physical world, making it seem as natural as using a regular physical object. The user no longer primarily focuses on the computer system; instead, it only reacts to the actions the user does in the actual world. In using physical objects as input devices, many requirements must be met.

- Accuracy—both the props and the tracking must be sufficiently accurate.
- Response—low latency is a requirement for the system.
- Interaction—it is important to clearly outline the different kinds that could occur. The props must be exact replicas of the information or object they represent. Disparities between the printed prop and the original data must be identified for several applications. Also necessary is for the prop to be correctly registered with its digital counterpart. The system's latency needs to be within acceptable bounds, and the mechanism for tracking the object needs to be accurate enough for the application. Additional difficulties arise if the tangible item is intended to be used in an augmented reality system. Last but not least, the techniques for using the prop must be clearly stated and understandable.

Ideally, a printed object would correspond exactly to the original model. However, this is never the case, as the manufacturing processes have specific resolutions and tolerances. If the difference between the model and the prop becomes too large for the intended usage, it must be compensated for. Props can be printed with tracking markers or tracking sensor attachment points already in place, or they can be a pure representation of the model, with the tracking device being attached to a random location on the prop. In the latter case, it is necessary to determine where that location is and how the prop is oriented with respect to the tracking device, but even if the prop is created with a known attachment point, it could be necessary to calibrate the prop.

- Registration and calibration: The printed prop is probably not going to be an exact replica of the model. Calibration is required to ascertain these variations. The prop may be skewed, have a marginally altered aspect ratio, or be of a different size. It's also possible that the prop be thicker or thinner than the original model; this shows up as an offset along the normal vector of the prop's surface and is caused by the material's contraction or expansion, certain aspects of the printing process, or some kinds of post-processing to follow printing of the prop (e.g. sanding, painting). Not only is it crucial to understand the precise geometry of the prop, but also its place-ment and orientation in relation to the tracking system. In an augmented reality system, this enables accurate registration of the actual view of the prop with the generated overlay as well as precise positioning of the associ-ated representation.

- Tracking system accuracy: Even a printed prop that matches exactly will be useless if the tracking system is not accurate enough. The needed track-ing system precision is influenced by the intended use as well as the size of the printed prop. The tracking for augmented reality systems must be at least one pixel accurate to register the prop with the augmentation. The precision must be adequate to pinpoint precisely the location on the puppet where the interactions occur if the prop is used in conjunction with a second input device that is utilised to perform interactions with the prop, such as indicating a location. This in turn is dependent on the size of the prop, the size of the smallest object, and the resolution of the original data. Figure 8.2 shows the refinement process using physical tools. A rudimentary model is built, and this model is used to create a physical prop. Then, using this prop as an input, a more accurate model is produced. These actions can be repeated numerous times. For instance, it might be sufficient if the system can identify the atom or relationship between atoms that is highlighted on the display in a molecular teaching programme. However, a precision mea-surement application might require one millimetre of accuracy. An accu-racy of less than 1 mm is required to precisely recognise collisions between a prop and a stylus in a two-handed arrangement. The tracked volume can be distorted by tracking technologies. If this distortion is static, it can be quantified and an error-correcting map can be made.

- Latency: There will invariably be a lag between the user's initial hap-tic feedback while manipulating a prop and the system's visual reaction. When there is lots of latency, the user performs worse. There is a visual lag between the movement of the physical props and the movement of the virtual additions in augmented reality systems where the user has a direct view of the interaction area (i.e. not video-based). On the other hand, in video-based augmented reality systems, there is no latency between the real and augmented components of the scene, only between the instant hap-tic feedback from the prop and the equivalent visual feedback from the enhanced video. Numerous studies have demonstrated that Weber's law is not followed when detecting the existence and quantity of delay. In a sys-tem with latency, this indicates that the least amount of additional latency a user can notice is not a fixed percentage of the base latency, but rather a constant quantity. This shows how individuals adjust to and make up for system latency. In order to maintain accuracy, the user must move more slowly as a result of this adjustment. According to research, consumers can detect a visual latency of 16 milliseconds (ms), and performance suffers when the visual system experiences a 50 ms latency. User performance is already impacted by haptic feedback latency at 25 milliseconds. The brain is able to disregard the more disrupted feedback and concentrate on the faster feedback when completing simple tasks that only demand one kind of stimulus (visual or haptic). Since there is no haptic or visual delay from the "representations" of the input devices when manipulating the props, using tangible props for interaction reduces the consequences of latency. The only latency arises from the time it takes for the system to react to an interaction event. In light of this, direct interaction should be planned to reduce the amount of time that users rely on computer input to complete tasks.

- Combined reality: Systems that use augmented reality blend computer visuals with the actual environment. There are numerous techniques to integrate the real and virtual scene, each of which has unique problems.

- Video cameras: They are used in video-based augmented reality to record the real world and add visualisations to the video stream. When a person views this video feed, they can see both real and virtual enhancements. To achieve co-location and mobility, a handheld device or a head-mounted display (HMD) might be employed. The real and virtual components of the scene are perfectly synchronised, but there will obviously be a delay between a user action and the associated visual response.

- Transparent or mirror-like display: In this scenario, the user's head and hands are placed in the interaction area between two semi-transparent mirrors. The mirror on which the augmentation is made reflects a display that is suspended somewhere above it. The distance between the mirror and the centre of the interaction area behind it is selected to be equal to that between the display and the mirror. This has the advantage of matching the focus distances of the user's hands or other objects in the interaction area with the virtual scene. For the real and the virtual scenes to be properly registered, head tracking is required. Although there is no delay from the real scene, motion temporarily throws the real and virtual scenes out of alignment. There are also see-through head-mounted displays (HMDs), which include displays and mirrors that are semi-transparent and use sophisticated optics to alter the perceived focal distance of the virtual scene.

- Projection: Texture augmentations are projected onto real things or even entire rooms of objects using one or more projectors. Simple geometric shapes or extremely intricate and detailed models can be the targets of the augmentation. The projection is changed to follow the object as it is moved about if the objects are both mobile and trackable. The augmentation can be made visible from all directions by using multiple projectors. It is feasible to add dimension to the enhancement by using stereo projecting and head tracking, as by having organs appear from the inside of a dummy [3, 11].

SURFACE MODELLING IN GENERAL

You have already seen a few examples of how basic geometric forms can be stretched, moved, or rotated to create surfaces. The optimal approach will depend on the specifics of the abstracted pattern you are attempting to create in three dimensions, but obviously this will apply through any number of configurations. It's conceivable that a little thought, experimenting, and (perhaps) simplification can go a long way. The visualization will be better when the parts used to build the surface are smaller, but the compilation, rendering, and generation of an STL file and eventually a G-code file will take longer. Minimizing support and making an artifact that will adhere to the work piece initially must also be taken into account, with the potential help of a raft or brim.

8.3. REVOLUTION IN MEDICAL FIELDS

The most popular and affordable type of technology for additive manufacturing is FDM. In this method, a heated print head is used to feed a thermoplastic filament, which is then laid down on the build platform layer by layer until the desired object is made. Some of the affordable desktop 3D printers that are commercially available are MakerBot, Flashforge, and Prusa. The diversity of materials that may be printed with these printers is a limitation, and lower-resolution objects are produced. There are other expensive FDM printers like Stratasys 3D printers that can print at greater resolutions and can use a wide range of materials. Because they can hold many print heads, FDM printers can simultaneously print a variety of materials. Typically, one of the print heads on these multi-head printers carries a supporting filament that is simple to remove or dissolve in water. Figure 8.7 shows the parts of an FDM 3D printer.

The most popular thermoplastic polymer utilised in the FDM method is ABS. Other frequently used printing filaments include PLA, polyamide, polycarbonate (PC), and polyvinyl alcohol (PVA). Because of their well-known biocompatibility and biodegradability, lactic acid-based polymers, particularly PLA and PCL, are widely employed in medical and pharmaceutical applications. Furthermore, PLA and PCL melt at low temperatures, 175 °C and 65 °C, respectively, which makes it simple to load pharmaceuticals without causing thermal deterioration that would reduce their bioactivity. In vivo, these polymers are hydrolyzed, and then they are

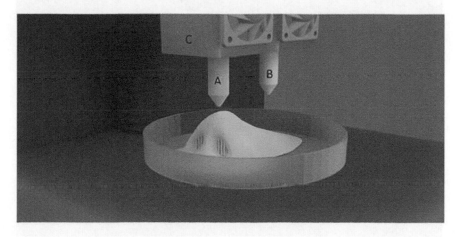

FIGURE 8.7 Dual Head FDM 3D Printer. (**A**) Building Material; (**B**) Supporting Material; (**C**) Print Heads [10].

excreted by excretory routes [12, 4]. Because PCL is less strong mechanically than PLA, it is employed in applications that do not require load carrying.

Hence it is used widely for prototyping in industry. In medicine, FDM is used for fabricating customized patient-specific medical devices, such as implants, prostheses, anatomical models, and surgical guides. Various thermoplastic polymers are doped with a variety of bioactive agents, including antibiotics, chemotherapeutics, hormones, nanoparticles, and other oral dosages, for personalized medicine. Using this technology, non-biocompatible materials, such as ABS or thermoplastic polyurethane (TPU), are used for creating medical models for perioperative surgical planning and simulations. These models are also used as a tool to explain the procedures to the patients before they undergo surgery.

8.3.1. Extrusion Based Bioprinting

Using either pressurized fluid or mechanical force, materials are extruded via a print head in this technique. As in FDM, layers of material are constantly applied until the desired shape is achieved, as depicted in Figure 8.7. This method is most frequently employed to create tissue engineering constructions with cells and anabolic steroids because it doesn't entail any heating techniques. The biomaterials utilised in 3D printing, known as "bioinks," are loaded with cells and many other biological components. Small forces of cells may be precisely deposited using this 3D printing technique, with little risk of process-related cell harm. Applications of this method in fabricating have significantly risen thanks to benefits such as precise cell deposition, control over cell dissemination rate, and process speed.

This method allows for the 3D printing of a wide variety of materials with various apparent viscosity and cells with high-density aggregates. Many different polymers are being studied for use in biomimetic technology. Bioinks for 3D printing are

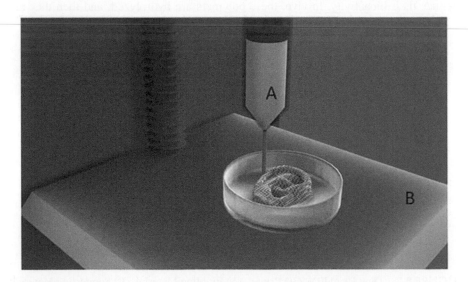

FIGURE 8.8   Extrusion Based Bioprinting. (**A**) Bioink; (**B**) Build Platform [10].

frequently made of natural polymers includes keratin, gelatin, alginate, and hyaluronic acid (HA) as well as synthetic polymers like PVA and polyethylene glycol (PEG). To improve the mechanical properties of the constructions, these bioinks are frequently post-processed using either chemical or UV crosslinking. Bioink can be used to create living tissue and scaffolds of varying complexity, depending on the kind of polymer applied. With this method, it is possible to print a complex cellular construct using many print heads containing various types of cell lines. In order to create a living human ear, Lee et al. used a six-head 3D printer with six distinct bioinks, including PEG as a sacrifice ink. Extrusion bioprinting has been employed by Laronda et al. to create gelatin-based ovarian implants that can hold follicles. The sterilised mice's ovarian functions were recovered by these implants, and they even produced offspring. Extrusion bioprinting has been employed to create scaffolds for the regeneration of several tissues, including bone, cartilage, the aortic valve, skeletal muscle, and neurons. Despite all of this achievement, the choice of material and mechanical strength is still a significant issue for bioprinting. This method still struggles with the difficulty of fabricating vascular permeability within a sophisticated tissue. Researchers have concentrated on using sacrifice materials to overcome this problem; these materials are integrated into the construct during 3D printing and removed during post-processing, leaving the blank spaces to serve as vasculature channels.

8.3.2. LIMITATIONS

Despite the reality that 3D printing can create complex patterns on demand and at low cost, its medical uses are constrained by the lack of variety in biomaterials. Medical 3D printing is still constrained by aspects including biomaterial printability, adequate mechanical strength, biodegradation, and biodegradable qualities even though a range of biomaterials are available, including metals, ceramics, polymers, and composites.

In order to achieve structural integrity of the final result in extrusion-based bioprinting, larger concentrations of polymers are typically used in the fabrication of bioinks. The scaffold's network layer and functional integration are hampered by this dense hydrogel environment. Vascularization is crucial for any moderately sized biological framework to perform, yet it is not feasible with present 3D printing technology. Researchers are currently printing small-scale scaffolds that can easily survive passive dissemination, but a livelihood functional organ requires extensive vascularization. Many researchers have employed the insertion of sacrificial elements during the scaffold creation to solve this issue. Following the fabrication of the structures, these components are eliminated using post-processing techniques. These materials cover the blank areas and give the printing materials mechanical support. Numerous sacrificial/fugitive materials, such as gelatin microparticles, pluronic glass, and carbohydrate glass, are now being studied [10]. Additionally, design-induced limitations cause material discontinuity, due to poor transformation of complex CAD design into machine instructions. Process-induced limitations include differences in porosities of CAD objects and finished 3D printed products [10, 12].

8.4. CHALLENGES OF DIGITAL CONSTRUCTION

Digital design greatly enhances efficiency in the construction phase by detecting and averting potential interference and by optimizing constructability. This phase also draws on other aspects of BIM and complementary technologies.

8.4.1. REAL-TIME DATA INTEGRATION, COORDINATION, AND SHARING AMONG STAKEHOLDERS

Providing the appropriate information at the appropriate time and location to all parties—including suppliers, subcontractors, and the on-site crew—during the building process is a significant problem. BIM in the cloud allows all stakeholders to combine and coordinate their actions, share data in real time, and exchange information. Collaboration becomes more effective and less prone to error in this way. The sports arena of the Washington Redskins in Washington, DC, is an intriguing example. As the columns and beams component was being built, BIM technology informed all supply chain partners that there were only about 100 requests for design clarification, in contrast to an estimated 10,000 for a project of a similar size that didn't use BIM.

8.4.2. DATA-DRIVEN CONSTRUCTION PLANNING AND LEAN EXECUTION

With the use of big data (derived, for example, from previous projects) and RFID surveillance, organizations may fine-tune their project-management tools for the best allocation of resources and programming at the construction site (of material, equipment, and labor). By doing this, businesses can keep a lean strategy while reducing non-value-adding tasks (including waiting, transferring workers, and hauling materials and equipment). For instance, the engineering, procurement, and construction firm Fluor uses a system dynamics model to simulate the impact of potential modifications on a variety of construction projects in order to improve decision making. This model is informed by previous project experience. When necessary, course corrections are performed, and significant savings—an estimated $700 million on a sample of 100 projects—are obtained. Virtual building models can support new methods of production, such as customization, preassembly, and 3D printing of specific components, by containing detailed information. There are numerous advantages, including greater material yields, reduced climate delays, safer work spaces, and better process sequencing for the construction industry. Training purposes and the identification of the most effective equipment handling techniques can both benefit from simulators of the assembly of complicated prefabricated assemblies.

8.4.3. AUTOMATED AND AUTONOMOUS CONSTRUCTION

At the construction site, robots and intelligent equipment increase output, accuracy, and safety. For tractors, large equipment, and dump trucks, for example, advanced amounts of automation (similar to computerised numerical control in industry) are made possible by remote control systems and 3D model guiding. The most difficult and dangerous jobs will eventually be given to fully or partially autonomous construction machinery. An example of this is the development of fully autonomous bulldozers by the Japanese equipment maker Komatsu. These machines are controlled by drones that monitor the surrounding region in real time and offer information about the amount of rocks and dirt that need to be moved.

8.4.4. RIGOROUS CONSTRUCTION MONITORING AND SURVEILLANCE

Companies can keep a closer eye on the processes and activities involved in building thanks to digital measurement and monitoring systems. 3D laser scanners are continuously used to compare constructions to the model in order to cut down corrective work. Additionally, telematics systems communicate data on various equipment metrics, such as fuel usage, for improved fleet management, while drones and remote cameras monitor building sites. The engineering department of Air Liquide, a leader in industrial gases, now takes complete 3D scans of its facilities, according to Benoit Potier, chairman and CEO of the firm. The use of robots in construction has been investigated since the early 80s [21]. Warszawski published one of the first critiques about the use of robots in the building sector and proposed different robot configurations to address different construction tasks [22].

For the technology to work with the materials used in 3D printing, numerous specifications must be met. Cementitious materials, polymer materials, and metallic materials are the three most frequently utilised materials in 3D printing, according to research. Fused deposition modelling (FDM) and fused filament fabrication (FFF) are processes that concentrate on printing things using polymer melting. Because polymeric materials lack structural qualities, they are typically employed for aesthetic purposes. The use of this material will offer a low-risk alternative for integrating additive manufacturing technologies into the building industry. Acrylonitrile butadiene styrene (ABS) and PLA are the two most widely used polymer base printing materials (polylactic acid). They are both thermoplastic polymers, which implies that when heated to a high temperature, they melt and then solidify again. Since PLA is biodegradable, it is thought to be more environmentally friendly than ABS. Construction materials with metallic qualities are also widely used, however 3D-printed constructions made entirely of metallic materials are quite heavy. The most common type of material employed in the aforementioned technologies, such as concrete printing and contour shaping, is cementitious material. It is best to utilise high-performance concrete for 3D printing. According to Le et al., high-performance concrete has been produced that may be used to construct both architectural and structural elements without the need for work piece.

8.5. BUILDING INFORMATION MODELLING (BIM)

As was already said, a 3D model is required in order to 3D print anything. Building Information Modelling (BIM) is the most widely used software platform in the construction sector. BIM is defined as the use of technologies of information and communication to the project lifecycle procedures in order to create a project that is safer, more effective, and more productive. The programme is used to create a design, and materials and project expenses are entered to produce an efficient project plan. The programme has already been included into conventional building projects, and it is already being used for design reasons. It is necessary to translate the CAD data acquired from the BIM programme into machine language. The most used format is called STL, which stands for stereolithography, the original term for 3D printing. The necessary value judgements for building a more sustainable infrastructure may be provided with the use of BIM. Sustainable design is utilised in building to advance sustainability. Sustainable design seeks to improve built environment quality while minimising adverse environmental effects. The use of BIM in construction automation has been studied; some of this research looked at the creation of new algorithms to automatically enter as-built structures into the BIM software. The construction process is made digital through the use of BIM in construction automation. Bentley Systems MicroStation and Autodesk Revit are two BIM programmes that are utilised in the Architecture, Engineering, and Construction (AEC) sector. The structural and mechanical characteristics of the concrete, such as its compressive strength and density, must be specified and accounted for in the BIM model in order for it to be used in 3D-printed construction projects. An increase in productivity, efficiency, and quality as well as a decrease in prices and lead times are all advantages of adopting BIM technology in 3D printing. Less design coordination mistakes, more energy-efficient design solutions, quicker cost prediction, and shorter manufacturing cycle times are some other benefits [3]. However, there are difficulties with the software's application in the building industry. Only 46% of respondents, according to research, believed that BIM had increased construction safety. This was brought on by the absence of BIM data, namely safety assessment information.

As with every revolutionary technology, 3D printing has a number of drawbacks. The impact on the current construction workforce is one of the social drawbacks. There won't be as many construction employees needed because of 3D printing. Even while this is viewed as advantageous since it lowers labour costs, people whose employment are put in jeopardy will suffer as a result. That could lead to societal issues in some areas that depend on building work. Another drawback that could be cited is the final product's quality. The printing procedure may have left the surface with a rough texture. Due to the design and material restrictions, 3D-printed structures might not meet end customers' expectations at the current level of technology. The 3D printer's chamber volume is a severe limit on how big the design may be. Larger-scale projects cannot yet be produced using 3D printing. Additionally, there are physical restrictions, where the printer is constrained by particular possibilities. The price is another drawback because 3D printing construction is now pricy. The initial cost of the equipment can be too high. The printer's transportation costs are difficult and pricey. The material being utilised is one of the drawbacks. A certain sort of concrete, which can be of lesser quality and more expensive, is needed for the concrete to be able to flow through the robot's nozzle. The lack of readily available appropriate material is viewed as a drawback. The drawbacks of requiring unique material for 3D printing are further examined in the section on material-related problems. The adoption and development of 3D printing in the construction sector are still in their infancy and confront a number of obstacles. A novel technique for 3D printing concrete material was created as a consequence of interdisciplinary study including material science, computing, and design. The difficulties may be divided into seven categories:

1. stakeholders,
2. material,
3. printer,
4. software,
5. computational issues,
6. architecture and design, and
7. construction management; and computational and software obstacles.

With the exception of the printer category, which is restricted to the extrusion-based approach, the issues are relevant to other 3D printing technologies.

8.6. PROBLEMS RELATING TO MATERIALS

This category encompasses issues with building materials including printability, buildability, and open time. So because needed concrete for 3D printing has different specifications than traditional building, new materials must be employed in construction to meet the demands of the new technology.

8.6.1. PRINTABILITY

The term "printability" describes a material's capacity to be pumped and printed. Pumpability and printability were defined by Lim et al. as the dependability and ease through which material may be delivered by a delivery system and a deposition device, respectively. To be pushed out of the nozzle of the 3D printer, the material must have the proper consistency. Whereas if material is just too hard, it will be difficult and energy-intensive to pump it through the pipe to the nozzle. If the material is too soft, the accuracy for inserting the material won't be correct and it would collapse quickly. Concrete printing is significantly influenced by the development of elastic properties in concrete and other important factors.

8.6.2. BUILDABILITY

According to Lim et al. [7], buildability is defined as the ability of deposited wet material to withstand deformation under stress. To be employed in 3D printing, the material must solidify quickly. In fact, if the substance takes a long time to solidify, it will collapse and lose its form. Additionally, 3D printing has strict criteria for building supplies. The rapid solidification of the materials is required by the rapidity of 3D printing. This demand cannot be satisfied by conventional building materials, necessitating extra R&D work. The concrete should also adhere to one another to create each layer, have sufficient buildability to allow it to lie down properly, maintain its place, and be strong enough to hold additional levels without cracking. As a result, as the concrete cures, it must support itself. The ability to realise particular designs is somewhat constrained since the material is extruded in a wet condition, requiring the build-up of layers to be done in a way that they are self-supporting in order to prevent collapse.

8.6.3. OPEN TIME

Open time was defined by Lim et al. [7] as the time frame during which printability and buildability are constant and within acceptable limits. Extrusion performs best when carried out continuously, but when stopped or restarted, it can lead to issues including under-printing (a halt in deposition that does not correspond with nozzle movement) and overprinting (depositing too little material). To get the material qualities required for additive manufacturing, special mixtures are often required.

The extensive list of references referenced shortly demonstrates the depth of research in the field of additive manufacturing, and particularly in the application of additive manufacturing to the construction industry. This chapter provides a thorough analysis of the literature on 3D printing in the building industry. It fills a gap in previous review articles of a similar nature and offers a more comprehensive discussion of management and technological issues. Twenty issues were identified and grouped into seven categories in the report. The six categories are used to identify and classify hazards. This fills a significant vacuum in the literature and offers fresh study avenues that concentrate on management as well as technical concerns. The article also reviews and assesses the various 3D printing techniques used in building [7, 8].

REFERENCES

1. Song, M. J. (2018). Learning to teach 3D printing in schools: How do teachers in Korea prepare to integrate 3D printing technology into classrooms? *Educational Media International*, *55*(3), 183–198. https://doi.org/10.1080/09523987.2018.1512448
2. Buehler, E., Easley, W., McDonald, S., Comrie, N., & Hurst, A. (2015). Inclusion and Education. In *Proceedings of the 17th international ACM SIGACCESS conference on computers' accessibility—ASSETS'15*. https://doi.org/10.1145/2700648.2809844
3. Lee, G., Nam, J., Han, H., & Kwon, S. (2019). A study on 3D file format for web-based scientific visualization. *International Journal of Advanced Culture Technology*, *7*(1), 243–247. https://doi.org/10.17703/IJACT.2019.7.1.243
4. García de Soto, B., Agustí-Juan, I., Hunhevicz, J., Joss, S., Graser, K., Habert, G., & Adey, B. T. (2018). Productivity of digital fabrication in construction: Cost and time analysis of a robotically built wall. *Automation in Construction*, *92*, 297–311. https://doi.org/10.1016/j.autcon.2018.04.004
5. 3D Printing in the Classroom—STEAM. (n.d.). https://sites.google.com/a/freeholdtwp.k12.nj.us/steam/home/resources/3d-printing
6. Ishii, H., & Ullmer, B. (1997). Tangible bits: Towards seamless interfaces between people, bits and atoms. In *CHI '97 proceedings of the SIGCHI conference on human factors in computing systems*, pp. 234–241.
7. Lim, S., Buswell, R. A., Le, T. T., Austin, S. A., Gibb, A. G., Thorpe, T. (2011). Developments in construction-scale additive manufacturing processes. *Automation in Construction*, *21*, 262–268. https://doi. org/10.1016/j.autcon.2011.06.010.
8. El-Sayegh, S., Romdhane, L., & Manjikian, S. (2020). A critical review of 3D printing in construction: Benefits, challenges, and risks. *Archives of Civil and Mechanical Engineering*, *20*(2). https://doi.org/10.1007/s43452-020-00038-w

9. Reynolds, R. (2015). Technology for teaching civics and citizenship: Insights from teacher education. *The Social Educator, 33*(1), 26–38.
10. Tappa, K., & Jammalamadaka, U. (2018). Novel biomaterials used in medical 3D printing techniques. *Journal of Functional Biomaterials, 9*(1), 17. https://doi.org/10.3390/jfb9010017
11. Kruszyński, K. J., & van Liere, R. (2009). Tangible props for scientific visualization: Concept, requirements, application. *Virtual Reality, 13*(4), 235–244. https://doi.org/10.1007/s10055-009-0126-1
12. Fang, F., Aabith, S., Homer-Vanniasinkam, S., & Tiwari, M. K. (2017). High-resolution 3D printing for healthcare underpinned by small-scale fluidics. *3D Printing in Medicine*, 167–206. https://doi.org/10.1016/b978-0-08-100717-4.00023-5
13. Conner, B. D., Snibbe, S. S., Herndon, K. P., Robbins, D. C., Zeleznik, R. C., & van Dam, A. (1992). Three-dimensional widgets. In *SI3D '92 proceedings of the 1992 symposium on interactive 3D graphics*, pp. 183–188.
14. Ishii, H. (2008). Tangible bits: Beyond pixels. In *TEI '08 proceedings of the 2nd international conference on tangible and embedded interaction.*
15. Hinckley, K., Pausch, R., Goble, J. C., & Kassell, N. F. (1994). Passive realworld interface props for neurosurgical visualization. In *CHI '94 proceedings of the SIGCHI conference on human factors in computing systems*, pp. 452–458.
16. Couture, N., Riviere, G., & Reuter, P. (2008). Geotui: A tangible user interface for geoscience. In *TEI '08 proceedings of the 2nd international conference on tangible and embedded interaction*, pp. 89–96.
17. Gillet, A., Sanner, M., Stoffler, D., Goodsell, D., & Olson, A. (2004). Augmented reality with tangible auto-fabricated models for molecular biology applications. In *VIS '04 proceedings of the conference on visualization 2004*, pp. 235–242.
18. Ortega, M., & Coquillart, S. (2005). Prop-based haptic interaction with colocation and immersion: An automotive application. In *HAVE 2005 proceedings of the IEEE international workshop on haptic audio visual environments and their applications, 2005.*
19. Kok, A. J., & van Liere, R. (2004). Co-location and tactile feedback for 2D widget manipulation. In *Proceedings IEEE conference on virtual reality 2004*, pp. 233–234.
20. Kruszynski, K. J., & van Liere, R. (2008). Tangible interaction for 3D widget manipulation in virtual environments. In *EGVE '08 proceedings eurographics symposium on virtual environments 2008*, pp. 89–95.
21. Haas, C., Skibniewski, M., & Budny, E. (1995). Robotics in civil engineering. *Computer-Aided Civil and Infrastructure Engineering, 10*, 371–381. https://doi.org/10.1111/j.1467-8667.1995.tb00298.x
22. Warszawski, A., & Rosenfeld, Y. (1994). Robot for interior-finishing works in building: Feasibility analysis. *Journal of Construction Engineering and Management, 120*(1), 132–151. https://doi.org/10.1061/(ASCE)0733-9364(1994)120:1(132)

9 Polymer 4D Printing— Future Perspective

9.1. INTRODUCTION OF 4D PRINTING

A device or item that can change from a 1D strand into a pre-programmed 3D shape, from a 2D surface into a pre-programmed 3D shape, or that can morph between multiple dimensions is said to be printed in four dimensions using a single substance or a combination of materials. The invention and advancement of 3D (three-dimensional) manufacturing technology is where the origin of 4D (four-dimensional) printing begins. Therefore, it is crucial to understand the evolution of 3D printing, which finally gave way to the development of 4D printing. The learner will comprehend the idea of 4D printing and its progress better if they have a basic understanding of 3D printing, including its history, printing processes, etc. Skylar Tibbits demonstrated how changes take place over time in a static produced item (via 3D printing) at the TED conference held at MIT in 2012. It has been shown that a straightforward 3D structure can develop into a more complicated structure over time. As a result, a new age of printing emerged that added a fourth dimension to 3D printing—time— and was dubbed 4D printing technology. In a straightforward manner, we may state that 4D printing is just 3D printing with the addition of a fourth dimension, time. Alternately, we may argue that the evolution of 3D printed components into 4D printed materials occurs throughout time. Following Tibbits' speech, academics and engineers from a variety of fields began to take notice of 4D printing. 2013 saw the publication of the first 4D printing research article (a year from the TED speech). One of the fundamental features of 4D printing is that it may change shape over time with the aid of a computer instruction that has been pre-programmed.

The difference between 4D and 3D printing: Imagine that 4D printing technology is 3D printing with the inclusion of time as a fourth dimension. The term "4D printing" is created by adding the concept of time to 3D printing. This makes it possible to pre-program items in various ways to respond to a variety of different inputs. Although 4D printing technology is futuristic, its prospects are quite promising. The ability to create any transformable shape that can be manufactured from a wide variety of materials is provided by 4D printing. These various materials will each have unique qualities as well as a variety of possible uses and applications. Developing dynamic self-assembling items that may change and be employed in a variety of industrial processes is a viable possibility [2].

The Gartner hype cycle must be reviewed if we want to examine the advancements of 4D printing in greater detail when compared with several other significant rising technologies globally. The Gartner hype cycle is an assessment of some of the most significant developing technologies globally and how they might fit with emerging business trends and applications over the next five to ten years. According

DOI: 10.1201/9781003349341-9

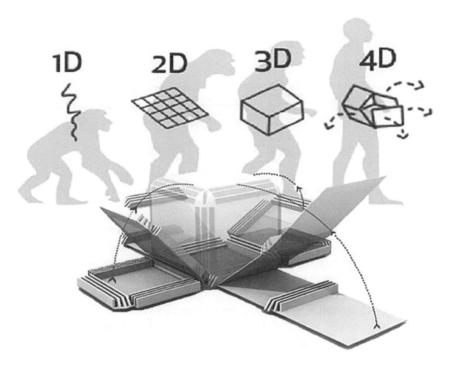

FIGURE 9.1 Fourth Dimension Geometric Structure [1].

to the hype cycle, 4D printing is currently at the point of trigger innovation. Over the next ten years, it will experience a rise in expectancy and progress and become a popular study topic. Looking at the articles and references of the research articles on 4D printing technology seems to be another technique to support this assertion about the hype cycle.

When opposed to 3D printing, 4D printing has a number of benefits, including the quick development of smart and multi-materials, more elastic and deformable shapes, and the ability to expand both 4D and 3D printing's possible application. The most current data on the subject shows an increase in 4D printing academic papers every year.

When compared to conventional approaches, it also delivers higher effectiveness, quality, and performance since 4D printed structures may self-improve their characteristics. The reduced material consumption of 4D printing aids in ensuring sustainable development. Printing of 4D models became simpler with the introduction of programmes like Kinematics (by Nervous Systems) and Project Cyborg (by Autodesk Research). These programmes allow the designer to see the object before it is printed, which makes it possible to fabricate 4D structures quickly [3].

9.2. FUNDAMENTALS OF 4D PRINTING

The structure created by 3D printing may evolve over time in regards to shape, property, and function. 4D printing is an enhanced progression of this technology.

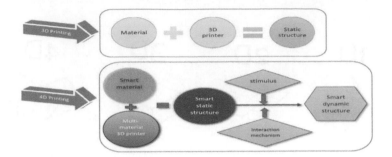

FIGURE 9.2   Block Diagram of 4D Printing [2].

Self-assembly, self-repair, and multifunctionality are all possible with 4D printing, which offers time-dependent, predictable, and fully programmable features. The idea of four-dimensional printing is primarily reliant on five elements: three-dimensional printers or other related machinery, stimulus-responsive materials, stimulation, inter-action mechanisms, and mathematical modelling, as shown in Figure 9.2.

Inkjet, fused deposition modelling, stereolithography, and laser sintering (sls) are now the most popular types of 4D printing. It is emphasised that the right mix of advanced textiles, including shape memory alloys, polymers, and smart nano-composites, is crucial for the dynamic operation of 4D printed items. The main technological advancement in 4D printing is smart material, which may change a pre-programmed process to react to particular external stimuli. Up to this point, 4D printing has improved the usage of smart materials. Shape memory compos-ites and polymers have been explored for their multi-functionality, which has been specifically mentioned in the aerospace industry. Moreover, shape memory hybrids and composites are gradually emerging, which has significant potential in industrial applications under the rapid growth in 4D printing. The slow development of shape memory hybrids and composites, which have enormous potential for industrial pur-poses under the explosive rise of 4D printing, is another development [3, 4].

9.3.   FACTORS RESPONSIBLE FOR 4D PRINTING

The five criteria seem to be mostly what 4D printing depends on, and while using 4D printing, all five aspects must be taken into consideration. These five elements are the additive manufacturing method, the printing medium, stimuli, interaction mechanisms, and modelling.

The printing method employed in the AM is the first component. Without the need for an intermediary tool, the AM technique permits the creation of printing material from digital information given by the computer. Numerous AM techniques exist, including stereolithography (SLA), selective laser sintering (SLS), fused depo-sition modelling (FDM), jet 3D printing (3DP), selective laser melting (SLM), direct ink writing (DIW), electron beam melting (EBM), etc. Almost all of these tech-niques can print a 4D substance so long as the printing medium is appropriate for the printer.

The second consideration is the printing medium, which needs to respond to stimuli while being sandwiched layer by layer. These elements are frequently referred to as smart materials or programmable materials (SMs). The sort of these smart materials that will be employed decides the stimuli that should be used, and how these materials react to the stimuli defines their self-transformation abilities.

The stimulus that will be employed when 4D printing is indeed the third element or aspect. Physical, chemical, and biological stimuli are all possible. Light, humidity, electromagnetic and electric energy, temperatures, ultraviolet light, and other physical stimuli are among them. Chemicals, pH, oxidant and reductant usage are examples of chemical stimuli. The biological stimuli include glucose and enzymes. When a stimulus is introduced, the structure experiences physical or chemical transitions, such as relaxation of stress, molecular motion, and phase shifts, which cause the structure to deform.

The interaction mechanism with mathematical modelling make up the fourth and fifth parts, respectively. Not all materials are capable of undergoing the necessary change when a stimulus is applied to smart material. We must offer an interaction method that will arrange the sequence of form change, such as mechanical loading or physical manipulation. The mathematical modelling is necessary to determine the period for which the stimulus will operate onto the smart material after supplying the possible interactions. 4D printing may be summarised as "The stimulus is put on the smart material utilising an appropriate interaction mechanism and mathematical modelling during an AM process which results in a 4D printed structure" in light of the variables that cause it [2].

9.4. LAWS OF 4D PRINTING

There are several resources and stimuli. As a result, the majority of current research in the field of 4D printing are scenario-specific. Any 4D printed structure has to have its time-dependent behaviour (the fourth D) anticipated. We were able to identify three universal rules that, despite the wide variety of materials and stimuli, control the "shape-shifting" behaviours of nearly all "multi-material" 4D printed structures through thorough yet methodical research of 4D printing and related fields.

9.4.1. First Law of 4D Printing

Nearly all of the "multi-material" 4D printed objects' shape-shifting behaviours (photochemical, photothermal, solvent, pH, moisture, electrochemical, electrothermal, ultrasound, enzyme, hydro, thermo, etc.) are caused by "relative expansion" between active and passive materials, a fundamental phenomenon. Nearly all complex 4D printing shape-shifting behaviours, including bending, tangling, curving, etc., are the result of this "relative expansion," which is made possible by encoding multiple kinds of anisotropy between active and passive materials and creating diverse heterogeneous substances.

9.4.2. Second Law of 4D Printing

Nearly all "multi-material" 4D printed objects exhibit shape-shifting behaviours, which are governed by four main forms of physics: mass diffusion, thermal expansion/

contraction, molecular transformation, and organic development. The comparative expansion among active and passive materials is caused by all of them (described and measured shortly), which also results in shape-shifting behaviours in response to stimuli (the stimulus is usually provided externally, but it can be internal) [5].

9.4.3. THIRD LAW OF 4D PRINTING

According to the third rule of 4D printing, "almost all multi-material 4D printed structures exhibit time-dependent form morphing behaviour that is controlled by two 'types' of time constants." Depending on the stimuli and material used for 4D printing, these factors may be equal, big, or disappear in relation to one another. Additionally, a mathematical bi-exponential equation for the fourth dimension was developed, which may be applied in the future to the modelling of 4D structures using hardware and software [2].

9.5. TECHNOLOGY FUNDAMENTALS OF 3D AND 4D PRINTING

9.5.1. FUNDAMENTALS OF 3D PRINTING TECHNOLOGY

The process of producing goods in a series of successive layers is known as additive manufacturing and it is a component of 3D printing. Depending on the input substance, 3D printing is primarily categorised as a solid, liquid, and power-based process. Fused deposition modelling (FDM) is utilised to create the solid-based pattern, while selective laser sintering (SLS) and selective laser melting (SLM) are employed to create the power-based pattern.

Almost 100 distinct types of desktop 3D printers, which are both compact and reasonably priced, are currently on the market. There are various forms of additive manufacturing, each of which has a particular set of input components and purposes. An overview is needed to fully comprehend its features and applicable approaches. In addition, a SWOT analysis of 3D printing technology shows its advantages, disadvantages, possibilities, and threats. The SWOT analysis of 3D printing using various methodologies is emphasised as being important to assess and point out its technical qualities.

9.5.2. FUNDAMENTALS OF 4D PRINTING TECHNOLOGY

The usage of 4D printing introduces the idea of change in the printed configuration over time, relying on external stimuli. 3D printing technology has been implemented to create static structures from digital information in 3D coordinates. The use of intelligent design and intelligent materials, which allow 4D printed objects to change in shape or functionality, is the main distinction between 3D and 4D printing. This suggests that any predicted time-dependent distortion of the framework should be completely encoded into the 4D printed structures. 4D printing is the creation of 3D printed objects with adaptive and programmable forms, qualities, or functioning as time dependent, according to a study group at the Massachusetts Institute of Technology (MIT).

Intelligent materials are able to recognise environmental stimuli and produce a helpful reaction. In this sense, intelligent materials may be thought of as those that offer a way to achieve an active intelligent reaction in a product that would otherwise be absent and have the ability to produce a wide range of improved capabilities and functions. For 4D printing to occur, three conditions must be met. The first method involves using stimuli-responsive composite materials that are mixed or contain multiple materials with different characteristics and are placed between layers. The second is the stimulus that will cause the item to move as a result of their interaction. These stimuli include, for instance, gravity, heat, cold, UV light, magnetic energy, wind, water, and even humidity. The last element is the duration of the simulation, and the outcome is a modification in the entity's state.

9.6. SMART MATERIALS

For 4D printing research to advance, smart materials are crucial. It's not necessary for intelligent materials to be able to alter form. Materials that can alter their hue, strength, or transparency are also crucial for biological applications, signalling, detecting alien objects, and camouflaging technology. A variety of stimuli can cause a time-dependent change in form, quality, or functioning. Gladman et al. activated 4D printed water-sensitive biomimetic structures that were based on natural forms using water. In addition to these stimuli, 4D printing may also make use of heat, pH, a combination of heat and water, along with light and temperature. Through controllable buckling, pneumatic transition, controlled wrinkling, photo-induced folding, stress-induced curing, heat controlled swelling, and the utilisation of shape memory polymers, the complex geometry may grow across many hundreds of orders in length scale [4, 6].

9.6.1. PROPERTIES OF SMART MATERIALS USED IN 4D PRINTING

- **Hydrogels:** Materials that are responsive to moisture. Matter that changes when it interacts with electrical energies.
- **Piezoelectric:** These react to applied mechanical stress, such as pressure or latent heat.
- **Thermo-reactive:** Heat or temperature fluctuations transform these materials.
- **Photo-reactive:** Materials catalyzed by light.
- **Magento-reactive:** Elements that transform when interacting with magnetic energy.
- **PH-reactive:** Matter that's triggered by pH levels.

9.7. 3D PRINTING VS 4D PRINTING

According to Xiao, four-dimensional printing is the next step up from three-dimensional printing, and 4D cannot exist without 3D. Establishing the foundations of both technologies is the first step in differentiating them. In a quick prototyping method also known as additive manufacturing, three-dimensional printing applies material layer by layer to construct three-dimensional things.

The 4D printing process uses the same technology to produce pieces. The extra step before hitting print is taken to establish the bonus dimension that sets one apart from the next in an object's geometric coding, as described previously. Depending on the asteroid's angles, measures, and dimensions, researchers encode the expected functioning of the thing in this case.

Therefore, 4D is about adding one more component to the dimensional composition, time, much as 3D printing is about adding depth to the constraints of a 2D construction—particularly, change through time. 3D-printed images retain their static, hard shape, but 4D-printed things will change into some type of motion.

In contrast to the more basic type of 3D printing, which can only produce items with a particular shape, four-dimensional printing is capable of producing objects that can alter shape and size after they've been printed.

9.8. 4D PRINTING EXAMPLES

Developments in 4D printing are presently only available for study and laboratory testing; they have not yet been commercialised. Any current uses, such as a breast implant that enables healthy tissue to grow in a cancer patient, are regarded as very experimental abnormalities that are awaiting rigorous testing and official permission before being widely used.

But it's crucial to remember that the foundation has already been built. Real-world applications for 3D printing will serve as a preview for what 4D printing will entail, at least for the foreseeable future. Researchers are immediately experimenting by introducing a self-activating twist to decades of 3D-printed proofs of ideas [7].

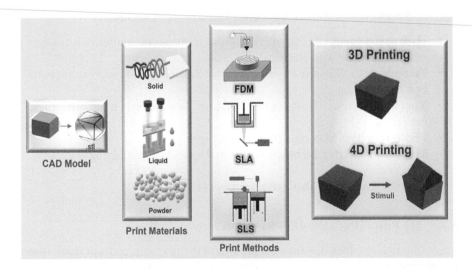

FIGURE 9.3   3D Printing Vs 4D Printing [2].

9.9. DIFFERENT TECHNIQUES OF 4D PRINTING

9.9.1. TESTING BY SKYLAR TIBBITS

The inventor of this technique, Skylar Tibbits, showed via his study that it is not only a pipe dream but will soon become a reality. A 2D grid layout that automatically changes shape by contracting across a 3D surface was created by Skylar Tibbits.

Materials with a single component that can swell to 100% of its original volume while the other component remains rigid. On the primary construction, expandable materials were inserted in appropriate locations to create joints that stretched and folded.

9.9.2. STRATASYS CONNEX TECHNOLOGY

While it may be challenging to implement this idea on a human scale, at least some businesses are reportedly intrigued, according to Skylar Tibbits. One such organisation that is supporting Skylar in his study is Stratasys. With this technique, any 1D strand may become a 3D shape, a 2D object can become a 3D object, or one object can change into another with only one print using a separate material's properties.

To trigger the self-assembly mechanism, researchers may use Skylar Tibbits' approach to programme various material characteristics into different particles of defined shape and achieve diverse water-absorbing qualities of materials. Water is used as the activation method here. With the use of this activation technique, programmability may now be added to non-electronic materials—for instance, robot-like activity independent of sophisticated electro-mechanical apparatus.

9.9.3. JERRY QI (ASSOCIATE PROFESSOR OF MECHANICAL ENGINEERING AT GEORGIA INT. OF TECHNOLOGY, USA)

One of the researchers striving to make this idea a reality is Jerry Qi. He has really no restrictions on his study because of his concept to use composites to evaluate whether this may work.

9.10. DEVELOPMENT OF 4D PRINTING

Composite materials now contain "shape memory" polymer fibres. Additionally, they created special "printed active composites" whose architecture include precisely where particular shape-memory fibres will behave differently in response to an external stimulus. Shape memory does not represent a novel idea; it is a self-assembling example where the thing remembers its shape and deforms in response to its environment. There are several examples of this type of self-assembly in addition to shape memory.

9.11. HOW QI'S 4D PRINTING TECHNOLOGY WORKS

For a number of composites, the team of researchers created specialised fibre structures at the laminate level. On the basis of the fibre design, size, alignment, spatial heterogeneity, and many other factors, composites with intriguing thermo-mechanical behaviours have been developed. The soft substances known as printed active composites (PACs) are made of glass polymer fibres that support an elastomeric matrix. When employed to make the active component of composites, these fibres are affected by the shape-memory phenomenon.

The PACs (printed active composites) at that time had thermomechanical programming that allowed them to accept 3D designs (shapes). The design of in-homogeneities at the micrometre level regulates the shape change. A CAD file was used to create the whole 3D design of the fibres and the matrix who used an object Connex 260 3D printer. The polymer ink droplets were then deposited at a temperature of around 70°C, wiped into a soft, smooth film, and then UV photopolymerized to create a layer that comprises matrix and fibre. The complete composite architecture was analysed after producing solitary lamina, multiple lamina, and finally the 3D laminate by printing numerous film layers.

Even while 3D printing is already playing a role in the emergence of a new revolution in the industrial sector by educating people about the technology, India does not yet have access to all of its potential applications. For instance, China is employing 3D printers to build buildings, and individuals in many other nations have begun to exploit this technology in virtually every industry imaginable. The idea of 3D bioprinting is already starting to take shape. By depositing substances, layer by layer, known as bio inks, tissues and organs are created. Recently, scaffold printing has also been incorporated into the process. You may utilise these scaffolds to rebuild ligaments and joints. Buildings or construction-related components can be created using the construction 3D printing technology. This method of construction is employed because it provides high-quality, high-performing building elements that suit the needs of architects and engineers.

The notion of 4D printing may be adopted based on the circumstance of how the nature has changed after 3D printing once the current 3D printing technology has been fully utilised. The 3D printing process serves as the foundation for 4D printing. Using 3D printers, certain specific materials that would be needed for 4D printing might be produced. The difference between 4D printing and 3D printing is that 3D printed structures may be pre-programmed to self-assemble or self-destruct. And there are differences in the use of biomaterials, where it is possible to print bio parts that may later be implanted in human bodies and observe changes in structure and function without the use of outside help [8].

9.12. APPLICATIONS OF 4D PRINTING

This section will go through some of the scientific fields where using 4D material has been beneficial, so that the reader will be introduced to these many sectors and perhaps achieve some exceptional outcomes in the future. Even though 4D printing technology is still in its early phases of development, it has already attracted a lot of attention from several research and technological domains. In order to create 4D printed materials that might be useful in the sphere of defence and protection, the US army has indeed launched a programme at three separate institutions. As an illustration, they have created 4D printed coatings for defence vehicles that can adjust to various weather conditions and alter their form, a soldier's outfit made of 4D printed material that may change shape as it comes into touch with a sharp instrument; the ability of the self-assembling 4D printed materials to change the course of a conflict is remarkable.

The impact of 4D printed materials on people's daily lives is astounding. In consideration of customer requirements, printed objects may adjust to shifting environmental conditions (humidity, temperature, pressure, wetness, etc.). Additionally, as 4D printed components take up less volume, less room and labour will be needed for their transportation and assembly. Due to their capacity to function independently and repair themselves, the items can be employed to satisfy new demands after being used. The potential for their use in space-related initiatives is enormous. It is possible to 4D print the solar cells and antenna used in satellites and spacecraft. They are able to self-assemble in a space, which reduces labour costs, and they can also adjust to the changing conditions of the environment. A potential application for 4D printed structures in the medical sciences is that, through a tiny incision, 4D printed stents may be inserted into the body, and once they are in the correct place, they can be altered by external stimuli to provide the desired effects. A Nano 4D printed gadget may also be inserted into the body, where it will self-assemble once it reaches the necessary area and perform the needed operation. A large reaction to 4D printing has also been seen in the fields of actuators and robotics, where they may be utilised to develop technology. In robotics, 4D printed structures can take the role of costly components like motors, sensors, and hard-to-assemble materials. Additionally, the use of these materials will result in smaller robots. Self-healing hydrogels, self-healing pipes, and fixing building flaws are just a few of the potential uses for 4D printed products that are anticipated in the future. The fields of sensors and printed replacement organs are further applications for these printed materials.

9.12.1. MEDICAL APPLICATIONS

The use of the material transfer method known as "bioprinting" to create biological materials such as cells, tissues, chemicals, etc., in order to carry out biological tasks. The rise in organ transplants led to the development of the discipline of bioprinting. The lack of available transplanted organs and the limits of conventional tissue engineering eventually led to the creation of a new subset of tissue engineering known as 3D or 4D bioprinting. When compared to conventional tissue engineering, bioprinting has a number of benefits, including the ability to create tissues with a high density and construct huge tissue-engineered objects. As technology developed over time, 3D bioprinting became 4D bioprinting. The process of printing biocompatible elements in three dimensions that can change over time is known as four-dimensional bioprinting. Here, the term "evolution" refers to the change in a 3D structure's physical, chemical, and biological makeup, as well as its characteristics and shape. The notion of 4D bioprinting has been expanded to include the functional development and maturation of 3D cells or tissues over time as a result of biological advancements. Therefore, 4D bioprinting involves the maturation of a three-dimensional structure employing a biomaterial (that may change form).

The key justification for 4D bioprinting's designation as the upcoming tissue engineering approach is its potential for building complex structures. Additionally, 4D bioprinting offers the best resolution of any bioprinting method, allowing for the incorporation of more data and specifics into the tissue. 4D bioprinting involves concentrating a laser pulse on a cartridge to remove material, which is then put on the substrate layer by layer. Materials suitable for 4D bioprinting must have the same responsiveness to temperature, humidity, light, electric and magnetic energy, stress, pH, and other factors as those previously addressed in the study. With the help of 4D bioprinting, it is now possible to create biological structures that can change shape in response to stimuli. Kuribayashi-Shigetomi et al. created a self-folding cube-shaped origami cell employing cell-filled 3D microstructures. Additionally, Kang et al. claimed the existence of an integrated tissue-organ printer (ITOP) that can produce stable human tissue in any form. Skeletal muscle, calvarial bone, and cartilage might all be produced by the ITOP. 4D bioprinting has shown to have enormous promise in biological applications, including medication delivery, tissue creation, and appropriate organ regeneration and transplantation. Furthermore, 4D bioprinting is becoming a more active area of study because of the creation of biomaterials that are completely compatible with 4D printing, such as printing scaffolds that can sustain multipotent human bone marrow mesenchymal stem cells (HMSC) using soybean epoxidized acrylate. The scaffolds produced in this manner were able to regain their former shape at 18.8 °C, the temperature of a human body. Additionally, the material demonstrated the same level of adhesion as PLA and PCL while outperforming PEDGA and other HMSCs. A 3D printed tissue was created utilising PCL that could alter form under resorption circumstances and tissue development over time to treat newborns with trachea bronchomalacia. The printed structure was made specifically for newborns under the age of one and it enlarges to accommodate the changing airway size over a three-year period. And after three years, the material started to disintegrate once the airways could work independently. The 4D influence of the PCL device's growth and degeneration over time suggests the possibility of 4D bioprinting tools that may be used to paediatric applications. Furthermore, a near-infrared (NIR) sensitive composite material was used to create a 4D printed model of the brain in order to study the behaviour of neural stem cells throughout 4D transitions. This is one of the sectors that is expanding as a result of the use of 4D printing, which allows us to create models of different organs to research their behaviour and makeup in order to develop better treatments for diseases that affect those organs. The use of biosensors, bio-actuators, biorobots, and other 4D printed nano-biomedical devices to track the physical effects in both cells and tissues has many potential uses.

The biorobots may carry out surgery as well as distribute therapeutics and medications. By utilising their self-deformation, 4D printing may be highly effective in curing aberration in the area of ortho-predicts. Additionally, the generated ortho-predict models have the ability to alter their characteristics in response to applied inputs. The ortho-predict parts created using 4D printing are much more adaptable, can be a precise reproduction of the patient's body, and can be used effectively in difficult procedures. In the upcoming years, 4D printing has a great deal of potential to grow in the field of orthopedics, where it may enhance the components utilised and operations performed. Modern 4D printing of advanced textiles may make it feasible to print organs like the heart, liver, and kidney since they can be flexible and self-compatible as well as being biologically more suitable to the individual. The skin transplant, which can be created via 4D printing and will match the patient's natural skin colour, can be utilised to treat skin burns that have the potential to heal like normal skin. When medication is needed at a particular spot, 4D printed devices can be employed to administer the treatment. When the surroundings at the chosen site are adequate for serving as stimuli, they can contain the medication and release it. Stents may be quickly and precisely manufactured with 4D printing. Because 4D printed structures have form memory, they need less surgical invasion and will resume their original shape after being implanted. SMP fibres were used as stents in a demonstration by Bodaghi et al. Ge et al. have created 4D printed stents utilising high-resolution SLA. 4D stents that respond to magnetic energy while also being regulated and directed remotely to the application region were created using DIW printing with PLA inks. The form and structure of scaffolds are now repeatable and consistent thanks to the use of 4D printing. As a result, 4D printing has enormous prospects in the development of medicine, which has the capability to enhance clinical outcomes, reduce surgical complications, fabricate scaffolds and stents for dental applications, deliver targeted drug delivery, as well as provide precise information more about anatomy of the body.

9.12.2. APPLICATION IN SOFT ROBOTICS

Hard materials like metals, tough polymers, and ceramics are frequently utilised to create robots in classical robotics. Such robots are made for certain environments and uses, and they are not resistant to any environmental factors. They are unable to achieve significant deformations and cannot carry out flexible duties. To get around these restrictions, a branch known as soft robotics was created, which can create robots that are flexible like people, can change how rigid they are, and can adapt to their surroundings. Soft and intelligent materials like electroactive polymers (EAPs) are needed for soft robots. These materials allow for a more gentle contact with delicate things than is possible with typical robots, giving them a higher tolerance for destructive pressures. Due to its flexibility, ability to bend, and ability to adapt to environmental changes, 4D printed structures are ideally suited for the field of soft robotics. Utilizing 4D printed structures, delicate robot actuators may be created. Dielectric elastomer actuators (DEAs) were traditionally made by hand, which created a time and labour issue. Rossiter et al. used 3D printing methods to create DEAs, which alleviated the issue. The DEAs are a type of smart material that can react to electrical stimuli, which is a subset of the EAP (4D materials). The actuator distorts as a result of the mechanical power that is created when electric energy is provided to DEA, which in turn causes the robot to move. As the study was carried out, the team created a sophisticated actuator model in which electrodes were affixed to both sides of every membrane; additionally, the actuator's parts were printed on two membranes in a pre-strain condition. When the top membrane received the voltage, the actuator moved higher, while when the lowering membrane received the voltage, the actuator went downward. By embedding pieces into unique components throughout 3D printing, it is possible to use 4D printing in soft robotics to enable parts to self-assemble after the product has been created. The inability to print one fully completed DEA at a time is another drawback to this approach. However, work may be carried out in this field in the future such that the frameworks won't need pre-straining and will be helpful in soft robotics. Soft actuators (SA) are inexpensive, simple to make, and environmentally benign, and they can be beneficial in soft robotics as well.

A composite material that may be utilised as SA was created by Miriyev et al. The SA can be easily manufactured using 3D printers since it was composed of a porous silicon elastomer matrix with ethanol-filled pores. The SA can pull weights and withstand stress and strain since it expanded when subjected to heat or warmth, which is caused by the ethanol evaporating. The SMPs have also been used as actuators (shape change), although robotics hasn't used them extensively because of their poor reaction. But as 4D printing technology advances, it is now possible to create intricate structures containing SMPs that may be activated by a variety of triggers. In particular circumstances, they can serve as actuators. As a possible use in optics, Lopez et al. produced a 4D structure utilising liquid crystal polymer (thermos-responsive). A 3D hydraulic robot was created by MacCurdy et al. utilising inkjet printing and non-curing fluids. Pneumatic artificial muscles (PAMs), which are light, elastic, extensible, and capable of producing significant forces, also were employed as actuators. Soft pneumatic actuators that may be utilised in soft robotics were printed by Yap et al. These were produced utilising FDM printing, resulting in robust constructions and more dependable and repeatable actuation. Despite significant advancements and progress in the field of 4D-printed soft robotics, modelling and complete control of these robots remain challenging tasks. This is because these robots have nonlinear dynamic behaviour, physicochemical

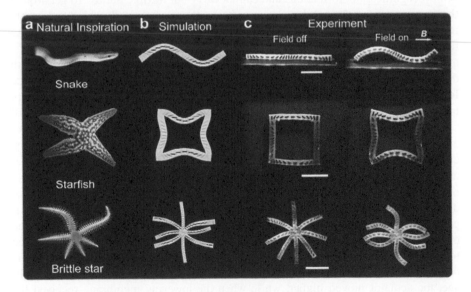

FIGURE 9.4   4D Printing Application in Soft Robotics [9].

characteristics that can change over time, and material rigidity. A closed-loop 4D printed soft robotics has been developed to provide improved control of the actuation. However, this is still in its early phases of development, and no significant research in this area has yet to be published. As a result, 4D printing has enormous promise for use in soft robotics, but additional materials of this type will need to be created in the future to enable improved actuation and control. Another issue that has to be resolved is the actuation efficiency and robustness of 4D printed robotic systems.

9.12.3. APPLICATION AS SELF-EVOLVING STRUCTURES

Self-evolving materials that can regain their previous shape upon contact with water may be created using 4D printing. These were initially created by Raviv et al. utilising various multi-materials, and the study team also gave them names. A hydrophilic polymer was used to print the structure on a plastic substrate; when exposed to water, the polymer would expand in proportion. The group created several intricate structures that displayed deformations including folding and stretching. The structures' mechanical degeneration was brought on by the constant wetting and drying of the structures. However, these issues can soon be resolved by investigating novel multi-materials having self-evolving properties. Additionally, additional causes, like temperature and UV exposure, can be investigated for such materials in addition to moisture. These self-evolving structures have several applications.

FIGURE 9.5   Self-Evolving Structures [10].

For instance, these constructions can be put up without difficulty in deep water, which is difficult to achieve with conventional structures. Additionally, this might be useful while repairing anything (ships, boats, etc.) in deep water. Instruments with such printable pieces that can be put together in a deep-water environment can also enhance the exploration of deep-water bodies and plants. Such self-evolving 4D printed structures can also be utilised to improve the diving suit used by rescuers in deep water.

9.12.4. APPLICATION IN AEROSPACE

The production of components for space missions is crucial to the aerospace sector. The components must be made for less money and have a long lifespan. As was already said, 4D printed buildings can survive harsh climatic conditions and are made at a reasonable cost. Additionally, these structures are adaptable to their environment. Thus, the aerospace sector is discovering possible uses for 4D printing. The possibility of using such materials in future space missions has been suggested by the International Space Station (ISS3D)'s printing of ABS constructions in microgravity. Additionally, they may be produced inside the spaceship and have a short printing time, which lessens the need for components from Earth. The utilisation of self-sustaining materials constitutes one of the mission's main components. The 4D structures can influence the course of such operations since they have the capacity to self-assemble in accordance with the circumstances. The shape memory alloy nitinol, which is composed of nickel and titanium, is extremely well liked in the space sector and has been created utilising 4D printing methods. The satellites, self-sustaining tools for space technology, can be made in part from 4D materials.

FIGURE 9.6 Aerospace Applications in 4D Technology [11].

A component's bulk is reduced by 80% because of the lightweight nature of 4D printed structures. PEEK, a thermoplastic that can withstand high temperatures and stresses with ease, has been printed using the SLS technology and is utilised to make parts for rockets and spaceships.

9.12.5. APPLICATION IN SENSORS AND FLEXIBLE ELECTRONICS

The creation of sensors can benefit from 4D printing, due to the 4D printed structures' responsiveness to factors including pH, humidity, temperature, pressure, and strain. For certain stimuli, they can therefore serve as sensors. Additionally, 4D printing is economical and produces lightweight, sensitive, accurate, and highly responsive structures. As a consequence, 4D printing will develop and modernise sensors. SMAs may be used to assess temperature, strain, and detect wear and degradation on the inside of a construction in addition to acting as actuators. Intelligent cars and planes may be produced using ultrasonic additive manufacturing (UAM), which mixes metals and smart materials to print intelligent structures. UAM is a sort of 3D printing that may be used to manufacture 4D structures and smart materials. UAM-printed structures have less bulk, which is advantageous for the transportation sector. Construction mistakes can be fixed with 4D printing since the parts may be shipped to the defective area and self-assemble there with the application of the right stimulus. Self-healing materials can be created by utilising the self-repairing capabilities of 4D printed structures. The durability and dependability of the material systems will be improved by the self-healing materials. Recently, structures that can self-heal cracks that form during 3D printing have been created. Because 4D buildings have the potential to repair themselves, they may be reused (recycled), which reduces material consumption and protects the environment from waste. In the textile sector, 4D printing may be used to create textiles that respond to shifting ambient circumstances and change as a result, improving comfort and ventilation. The colour and texture of 4D printed garments may also alter in response to environmental changes. Materials, notably PLA, ABS, and Nylon, might be printed mostly on surfaces of textiles including cotton, polypropylene, poly-wood, and polyester. Additionally, 3D printed clothing has begun to acquire popularity in the film and fashion industries. In addition, 4D printing may improve the quality of household appliances, making things more cosy and moisture- and heat-resistant. In response to conditions and surroundings, 4D printed shoes may change. An airplane's 4D printed wings can adjust to various air conditions and offer improved lift and protection. It has been established that 4D printed composite springs are functionally equal to conventional springs when utilised as leaf springs in vehicles. The wind turbine blades were manufactured using 4D printing without the need for typical electromechanical devices like sensors and actuators. These blades' 4D printing increased system and energy production control as well. The same research team also developed a smart photovoltaic system that can change shape in response to sunshine availability, resulting in an effective solar energy production system. One further use of 4D printing is to create electronic devices and circuits [2, 12–16].

FIGURE 9.7   Application of Flexible Electronics [17].

9.13.   ADVANTAGES OF 4D PRINTING

The ability to self-modify in 4D printing allows it to alter form, contract, and expand. The use of 4D printing allows for the compression of bigger things. Designing a heart that can change form when given the right material is helpful. According to the need, 4D printing may change its shape and functionality. This method accurately produces heart valves with no restrictions on geometry. Heart, kidney, and liver can all be printed using intelligent material. It is possible to create intelligent valves that can regulate blood circulation by changing the diameter of the gate. This method also modifies the setup in real time for all 3D printing applications. With the passage of time and variations in temperature, objects' shapes alter. Up to 90% of the labour and materials may be saved by using 4D printing [15, 16].

9.14.   DISADVANTAGES OF 4D PRINTING

When compared to 5D printing, the modulus of 4D printing is quite low. Regarding the temperature of the surroundings, they are less steady. In 4D printing equipment, the printer head's smart material loading is challenging. When compared to 5D printing, 4D manufactured components are weaker. It is unable to produce integrated pieces with complicated structures including curved surfaces. Therefore, 5D printing has been developed to counteract these detrimental effects [16].

9.15.   MOTIVATIONS

A structure may be triggered for self-assembly, reconfiguration, and replication using ambient free energies thanks to 4D printing, which opens up completely new domains of application. This technique offers a number of benefits, including the ability to change flat-pack 4D-printed components and a large volume decrease for storage. Another illustration is the ability to 3D print basic components made of intelligent materials initially, and then allow them to self-assemble into the final complicated shape, as opposed to directly manufacturing complicated structures. The three major areas of 4D printing's prospective uses—self-assembly, multi-functionality, and self-repair—can be used to group together most of these uses. The creation of minimally invasive surgical equipment that can be inserted into a person's body through a small surgical incision and afterwards assembled at the necessary position for surgical operations is one example of how the potential of 4D printed structures to self-assemble and self-repair emerges great possibilities for implementation.

9.15.1. SELF-ASSEMBLY

Future applications could take place in hostile settings and on a huge scale. Small 3D printers may be used to manufacture individual components that can later be self-assembled into bigger constructions like satellites and space antennas. This expertise may be used to design sophisticated part transportation systems for the international space station. Additional uses include self-assembling structures, which are particularly helpful in conflict zones or in outer space where the components may join together to produce a fully formed structure with minimal labour force. The usage of 4D printing also has the additional benefit of removing some building limits. To make particular sections of a product that serve as links and hinges for bending, rigid materials and smart materials may both be 3D printed. According to Raviv and colleagues, construction must become more intelligent in order to address issues with wasteful use of time, resources, money, and energy. These problems may be resolved by incorporating information into the materials utilising design tools and programmes, which will improve the accuracy of the materials and construction. Self-assembly might not be effective for all uses, therefore it's important to identify the industries and uses that stand to gain the most from self-assembly.

9.15.2. SELF-ADAPTABILITY

With the use of 4D printing, sensing and movement may be integrated directly into a material, eliminating the need for additional electromechanical devices. This reduces the quantity of electromechanical system-related failure-prone devices, assembly time, material and energy expenses, and the number of components in a structure. This technique is being used to create self-adaptive 4D printed tissues and 4D printed customised medical equipment like trachcal stems.

9.15.3. SELF-REPAIR

The capacity of 4D printed objects to fix errors and repair themselves offers significant benefits in terms of reuse and recycling. Some possible uses for 4D printing include self-healing hydrogels and self-healing pipes. Several types of reactions, such as covalent bonding, supramolecular chemistry, H-bonding, ionic interactions, and stacking, can cause polymers to repair on their own. The capacity of self-healing materials to make repairs ranging from bulk fractures to surface scratching has been proven to offer considerable potential for developing soft actuators with greater durability. Self-healing hydrogels have been effectively used as inks in rapid prototyping [4].

9.16. FUTURE DIRECTIONS AND OPPORTUNITIES

According to a market analysis published on the Reuters website, the market for 3D printing will increase at a CAGR of 30.20% from 2017 to 2022, while the market for 4D printing may expand at a CAGR of 40.30%. Without a doubt, the overall spending in 3D printing is currently more than that in 4D printing. The aforementioned figures contrast the expansion (the projected future) of 3D and 4D printing, nevertheless. By itself, 3D printing is regarded as a multi-disciplinary activity. As a result, additional research fields will be involved in the field of 4D printing. The power of 4D printing may be increased by this kind. Future works may take a variety of angles. The interoperability of materials in multi-material constructions is one of the primary themes. At their contact, the materials ought to solidify into a bond. Their relationship ought to be resilient in the face of stress. The other subject is the measurement and modelling of some of the time-constants and other bi-exponential equation factors. For active–passive materials, a few of the variables need to be evaluated, while designs (whether generic or case-specific) can be created for some of them. For different types of materials, experimental research can be used to determine the precise values or factor ranges. The next subject is the "integration" of hardware and software improvements. Software for 4D printing in the future should be able to forecast certain outcomes. Additionally, it can offer a scenario for gradually fine-tuning the behaviour. Future hardware advances for 4D printing will need some multi-material printing-capable control methods. Imagine that a solitary construction will encode ten distinct materials simultaneously through ten different nozzles at each of the structure's numerous positions (voxels). The following section focuses on the printability of smart materials. There are various intelligent materials, however they must be made printable. Another key subject is fine-tuning the reaction time. The following point relates to the creation of products using 4D printing. In addition to being a concept, the manufacturing paradigm of 4D printing connects design and production in a way that results in a finished product. Therefore, it is important to always think about and approach new goods or applications that might benefit from 4D printing's distinctive properties.

A number of 4D printing-specific research and development initiatives are under progress in a number of sectors, including healthcare, electronics, automotive, aerospace and defence, consumer appliances (fashion and consumer durables), cloth, architecture, and industrial equipment. Despite being a new technology, many industry experts agree that 4D printing has a wide range of potential applications.

- The field of additive manufacturing is still in the early stages of development. New and better machinery, printing techniques, software, and materials are always being created. Because 4D printed structures have the capacity to change throughout time in shape or functionality in response to stimuli like pressure, temperature, wind, water, and light, 4D printing has increasingly attracted interest. The use of intelligent materials, designs that anticipate process changes, and smart printing in a variety of fields—from straightforward form modifications to printing living organisms—are all part of 4D printing technology. 4D printing has been created using multi-material

3D printing and intelligent materials. This brand-new technology offers a workable way to create a small deployable structure. The foundation of 4D printing is smart materials.

- Smart materials enable 3D-printed objects to self-assemble and change shape in the presence of external stimuli, cutting down on production time and costs.
- Self-healing polymers can also be used when smart materials are used. Pipes that develop leaks may be able to mend themselves with the help of self-healing hydrogels.
- Smart materials eliminate the need for additional electromechanical systems by performing sensing and actuation functions directly inside the substance. As a result, there are fewer structural components that need electronics and electromechanical actuators.

The shape-shifting characteristic of 4D printing, along with other polymer qualities like colour and texture, may be used for a variety of purposes. This might be helpful for smart fabrics that respond to diverse inputs from the environment and colour change and texture to effectively ventilate or insulate the wearer, improving convenience and usefulness [2, 4].

Due to extensive research and development efforts, the industry for 4D printing is starting to take shape. There is disagreement among experts on market expansion. According to the positive assessment of the technology, the market will expand at a CAGR of almost 33%, from US$35 million in 2019 to US$200 million in 2025. The 4D printing industry, however, is expected to develop at a somewhat slower pace of 20% by 2025 due to the fact that it is a unique technology still in its infancy, according to FutureBridge.

While being a revolutionary technique, 4D printing must yet clear a number of technical obstacles before it can be extensively used. The inability to produce support structures for complicated items, the absence of multi-material printers, the lack of affordable printers and smart materials, the lengthy print times, and the restricted long-term dependability of printed objects are only a few of the industry's key obstacles. While there have been some developments in printing technology, such as the use of five-axis printing apparatus, which is anticipated to solve the issue of creating support structures for intricate interior structures, other difficulties still exist.

Other barriers to the widespread use of 4D printing technology include lack of control over intermediate stages of contraction, sluggish and imprecise actuation, and restricted material supply. However, given the enthusiasm from manufacturers and the intense degree of research and development efforts surrounding 4D printing, the technology may advance exponentially at a rate that is quicker than anticipated. Furthermore, manufacturers need to keep up with technical developments and possible consequences of 4D printing if they wish to remain at the forefront of technological advances and breakthroughs [18].

REFERENCES

1. (2023, January 3). 4D Printing: All you need to know in 2023. *Sculpteo*. www.sculpteo.com/en/3d-learning-hub/best-articles-about-3d-printing/4d-printing-technology/

2. Abdussalam, A. A., Musbah, A., & Faraj, A. (2020). 4D printing technology: A revolution across manufacturing. *International Journal of Mechanical and Industrial Technology, 7*, 45–51.

3. Ahmed, A., Arya, S., Gupta, V., Furukawa, H., & Khosla, A. (2021). 4D printing: Fundamentals, materials, applications and challenges. *Polymer, 228*, 123926. https://doi. org/10.1016/j.polymer.2021.123926

4. Nkomo, N. (2018). A review of 4D printing technology and future trends. In *Eleventh South African conference on computational and applied mechanics SACAM 2018.*

5. Momeni, F., & Ni, J. (2018). Laws of 4D printing. *arXiv.* https://doi.org/10.48550/ arXiv.1810.10376

6. Haleem, A., Javaid, M., Singh, R. P., & Suman, R. (2021). Significant roles of 4D printing using smart materials in the field of manufacturing. *Advanced Industrial and Engineering Polymer Research, 4*(4), 301–311. https://doi.org/10.1016/j.aiepr.2021.05.001

7. Becher, B. (2022, November 1). What is 4D printing? *Built In.* https://builtin. com/3d-printing/4d-printing

8. Ramesh, S., Reddy, S. K., Usha, C., Naulakha, N. K., Adithyakumar, C. R., & Lohith Kumar Reddy, M. (2018). Advancements in the research of 4D printing–a review. *IOP Conference Series: Materials Science and Engineering, 376*, 012123.

9. Qi, S., Guo, H., Fu, J., Xie, Y., Zhu, M., & Yu, M. (2019). 3D printed shape-programmable magneto-active soft matter for biomimetic applications. *Composites Science and Technology, 188*, 107973. https://doi.org/10.1016/j.compscitech.2019.107973.

10. McAlpine, K. J. (2016, October 18). 4D-printed structure changes shape when placed in water. *Harvard Gazette.* https://news.harvard.edu/gazette/story/2016/01/4d-printed-structure-changes-shape-when-placed-in-water/

11. Carlota, V. (2019, April 3). 4D printing reconfigurable materials for use in aerospace, medical and robotics fields. *3Dnatives.* www.3dnatives.com/en/4d-printing-materials-030420195/

12. Rastogi, P., & Kandasubramanian, B. (2019). Breakthrough in the printing tactics for stimuli-responsive materials: 4D printing. *Chemical Engineering Journal, 366*, 264–304.

13. Wu, J.-J., Huang, L.-M., Zhao, Q., & Xie, T. (2017). 4D printing: History and recent progress. *Chinese Journal of Polymer Science, 36*, 563–575.

14. Javaid, M., & Haleem, A. (2019). 4D printing applications in medical field: A brief review. *Clinical Epidemiology and Global Health, 7*, 317–321.

15. Liu, T., Liu, L., Zeng, C., Liu, Y., & Jinsong, L. (2019). 4D printed anisotropic structures with tailored mechanical behaviors and shape memory effects. *Composites Science and Technology, 186*, 107935. https://doi.org/10.1016/j.compscitech.2019.107935

16. Zhang, Z., Demir, K. G., & Gu, G. X. (2019). Developments in 4D-printing: A review on current smart materials, technologies, and applications. *International Journal of Smart and Nano Materials, 10*(3), 205–224.

17. Industrial Research and Consultancy Centre. (n.d.). Flexible electronics: Electronics that can bend, flex, stretch or fold. https://rnd.iitb.ac.in/research-glimpse/flexible-electronics-electronics-can-bend-flex-stretch-or-fold

18. Prakash. (2020). 4D printing – The technology of the future. *FutureBridge.* https://www.futurebridge.com/industry/perspectives-mobility/4d-printing-the-technology-of-the-future/

Index

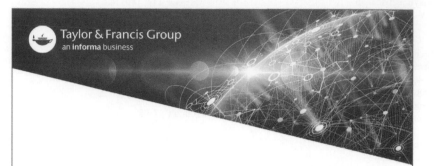